The Intelligent Movement Machine

The Intelligent Movement Machine

An Ethological Perspective on the Primate Motor System

MICHAEL S. A. GRAZIANO, Ph.D.

2009

OXFORD
UNIVERSITY PRESS

Oxford University Press, Inc., publishes works that further
Oxford University's objective of excellence
in research, scholarship, and education.

Oxford New York
Auckland Cape Town Dar es Salaam Hong Kong Karachi
Kuala Lumpur Madrid Melbourne Mexico City Nairobi
New Delhi Shanghai Taipei Toronto

With offices in
Argentina Austria Brazil Chile Czech Republic France Greece
Guatemala Hungary Italy Japan Poland Portugal Singapore
South Korea Switzerland Thailand Turkey Ukraine Vietnam

Copyright © 2009 by Oxford University Press, Inc.

Published by Oxford University Press, Inc.
198 Madison Avenue, New York, New York 10016
www.oup.com

Oxford is a registered trademark of Oxford University Press

All rights reserved. No part of this publication may be reproduced,
stored in a retrieval system, or transmitted, in any form or by any means,
electronic, mechanical, photocopying, recording, or otherwise,
without the prior permission of Oxford University Press.

Library of Congress Cataloging-in-Publication Data
Graziano, Michael S. A., 1967–
The intelligent movement machine : an ethological perspective on the primate motor system /
Michael S.A. Graziano.
p. ; cm.
Includes bibliographical references and index.
ISBN-13: 978-0-19-532670-3
1. Motor cortex—Physiology. 2. Human locomotion. 3. Animal locomotion.
4. Primates—Physiology. I. Title.
[DNLM: 1. Motor Cortex—physiology. 2. Movement—physiology. 3. Models, Animal.
4. Primates—physiology. WL 307 G785i2009]
QP383.G78 2009
612.8'252—dc22

2008012835

9 8 7 6 5 4 3 2 1

Printed in the United States of America
on acid-free paper

Acknowledgments

This book would not have been possible without many collaborators and colleagues including Charlie Gross, my mentor through many years; Charlotte Taylor and Tirin Moore, who collaborated on the initial experiments in which we electrically stimulated the monkey motor cortex on a behaviorally relevant time scale; Dylan Cooke who studied the motor cortex control of defensive movements; Tyson Aflalo who explored the computational and theoretical implications of the data; Sabine Kastner and Jeffrey Meier who made it possible for my lab to explore functional imaging in the human motor cortex; Tyler Clark who acquired many books and articles that went into the research; and Theodore Mole who remains an unfailing presence in the lab. In addition, Ed Tehovnik, Sabine Kastner, Charlie Gross, and Craig Panner made many useful comments on the manuscript.

Preface

In 1870, Fritsch and Hitzig discovered the motor cortex in the dog brain. Since then, for one hundred and thirty years, researchers have grappled with the fundamental question of motor cortex: How is it organized?

I believe this question is finally answered. The answer is simple in concept. An animal's normal movement repertoire is flattened onto the cortical surface. The complexity of the map comes from the complexity of the movement repertoire. With a good description of the typical movement repertoire of a species of animal, it should be possible to predict mathematically the layout of the motor cortex. We now have an approximate description of the movement repertoire of macaque monkeys, and with it we can explain the overarching organization of the monkey motor cortex.

The theory that the motor repertoire is flattened onto the motor cortex is one specific example of a general principle of brain organization. One might say that the mental repertoire of the animal is mapped somehow onto the entire brain. In the case of movement, the repertoire is conveniently observable and therefore its mapping onto the cortical surface can be studied directly. The purpose of this book is to review experiments on how the motor repertoire is mapped onto the cortex, ranging from the initial discovery of motor cortex to the present.

Contents

Chapter 1: Introduction *3*

Chapter 2: Early experiments on motor cortex *13*

Chapter 3: An integrative map of the body *39*

Chapter 4: Hierarchy in the cortical motor system *51*

Chapter 5: Neuronal control of movement *71*

Chapter 6: What can be learned from electrical stimulation? *85*

Chapter 7: Complex movements evoked by electrical stimulation of motor cortex *97*

Chapter 8: The match between natural neuronal properties and stimulation-evoked movement *125*

Chapter 9: The movement repertoire of monkeys *139*

Chapter 10: Dimensionality reduction as a theory of motor cortex organization *151*

Chapter 11: Feedback remapping and the cortical-spinal-muscular system *167*

Chapter 12: Social implications of motor control *181*

Literature Cited *199*

Index *219*

The Intelligent Movement Machine

Chapter 1

Introduction

BACK STORY: MIXING TWO EXPERIMENTAL CULTURES

When I was a postdoc at Princeton University, I worked on the integration of vision, touch, and movement in the monkey brain. The experiments involved monitoring the activity of single neurons in the motor cortex during the monkey's movements or during the presentation of sensory stimuli. I had never considered using electrical stimulation to study motor cortex. Many colleagues had suggested the technique, and my response was something like, "You get muscle twitches. Big deal. You can't really learn anything."

At the same time another postdoc in the lab, Tirin Moore, had begun a set of experiments on the frontal eye field of monkeys. He used a common electrical stimulation technique in which pulses of current are delivered into the cortex through a fine, hair-like electrode. The pulses are presented in a train at high frequency (typically 200 pulses per second). This method directly activates a small sphere of brain tissue around the electrode tip. The directly stimulated neurons then recruit physiologically connected networks. If stimulation is applied to a spot in the frontal eye field, it evokes an eye movement that closely resembles a natural one.

When experimenting on the frontal eye field, for each new monkey studied, one typically first explores a broad area of cortex, stimulating in a variety of locations to find the borders of the area of interest. During one such exploration, Tirin came running down the hall to my office, his lab coat billowing behind him like the cape of a superhero, and said, "Mike, you have to look at this." I came and looked.

He held a button in his hand, and every time he pressed the button, the monkey sitting in the plastic monkey chair in the center of the room extended his arm forward and shaped his fingers as if reaching for something invisible. The effect was immediate, consistent, and obviously as amazing to the monkey as it was to us because the monkey grabbed hold of his hand with the other one, pulled it straight down, and sat on it, effectively ending the experiment for the day. Tirin had obviously missed the frontal eye field and gotten the electrode into the primary or premotor cortex.

"We have *got* to study this," he said.

The evoked movement was no muscle twitch. The reason was immediately obvious to us. In a standard stimulation experiment on motor cortex, the stimulation is applied in a brief burst for 50 ms or less. The result of this brief stimulation is a muscle twitch. But little if any behavior unfolds on such a short time scale. Neurons in motor cortex are not normally active in 50 ms bursts but instead, to a first approximation, are active throughout the duration of a movement. In the present case, the stimulation was applied for half a second, approximating the duration of a monkey's reaching or grasping. As a result, instead of a muscle twitch, a complete movement unfolded.

After a month of mulling and of dinner conversations at the local Italian restaurant, three of us began the new experiment: Tirin, myself, and Charlotte Taylor, a graduate student also in the lab. We set out to study the motor cortex using the technique of stimulating on a behaviorally relevant time scale.

Our procedure was to sit for hours in front of the monkey like a panel of judges, studying one cortical site in a day, stimulating it hundreds of times under every condition we could think of, watching every event, discussing every detail, and arguing over exactly what description to write in the data book. In addition to our general contributions to the experiment, we each had specific duties. I was the scribe. Charlotte operated the button that delivered the stimulation to the cortical site. Tirin fed the monkey a constant supply of raisins to calm him and entice his arms into a variety of test configurations (The monkey eventually became obese from the constant snacking.). The study was intentionally as unstructured and as observational as possible. We did not know what to expect.

On the first day that we reached the motor cortex it became abundantly clear that stimulation evoked complex movements combining many joints. We were able to evoke integrated movements of the shoulder, arm, and hand. We also noticed that regardless of the starting position of the arm, the movement evoked by stimulation seemed to bring the hand toward the same final position as if in a goal-directed action.

A few days later we encountered a site in the cortex where stimulation caused the fingers to close in an apparent grip, the hand to move to the mouth, and the mouth to open. The monkey appeared to be feeding himself, even though there was nothing in his hand. The movement was so natural, so utterly like the monkey's normal feeding action, that triggering it by button push gave us the willies. It was uncanny. We ran out of the experiment room and searched the halls for someone, anyone, to look at the result and tell us that it was real, that we weren't nuts.

We wondered if the monkey was inadvertently fooling us. Perhaps the stimulation caused merely a general tendency to move and the monkey then supplied a movement that was on his mind, so to speak, because he was constantly feeding himself raisins. This explanation seemed unlikely because we evoked the hand-to-mouth movement only from one region of cortex, and the evoked movement had a mechanical reliability. However, we tested the possibility by injecting an anesthetic into the monkey and waiting until he was asleep. Stimulation of the same site in cortex still drove the fingers into a grip, the hand upward toward the mouth, and the mouth open. The movement had nothing to do with the monkey's behavioral context. It was as mechanical as clockwork. We appeared to have tapped into its control mechanism.

1. Introduction

As the experiment continued, we uncovered more actions that looked like they were straight out of the monkey's natural repertoire and that could be generated by stimulating specific sites in the motor cortex. Different zones within the motor cortex appeared to emphasize different major categories of action. Some of these action categories are illustrated in Figure 1-1. They included ethologically relevant behaviors such as closing the hand in a grip while bringing the hand to the mouth and opening the mouth; extending the hand away from the body with the grip opened as if in preparation to grasp an object; bringing the hand inward to a region just in front of the chest while shaping the fingers, as if to manipulate an object; squinting the facial muscles while turning the head sharply to one side and flinging up the arm, as if to

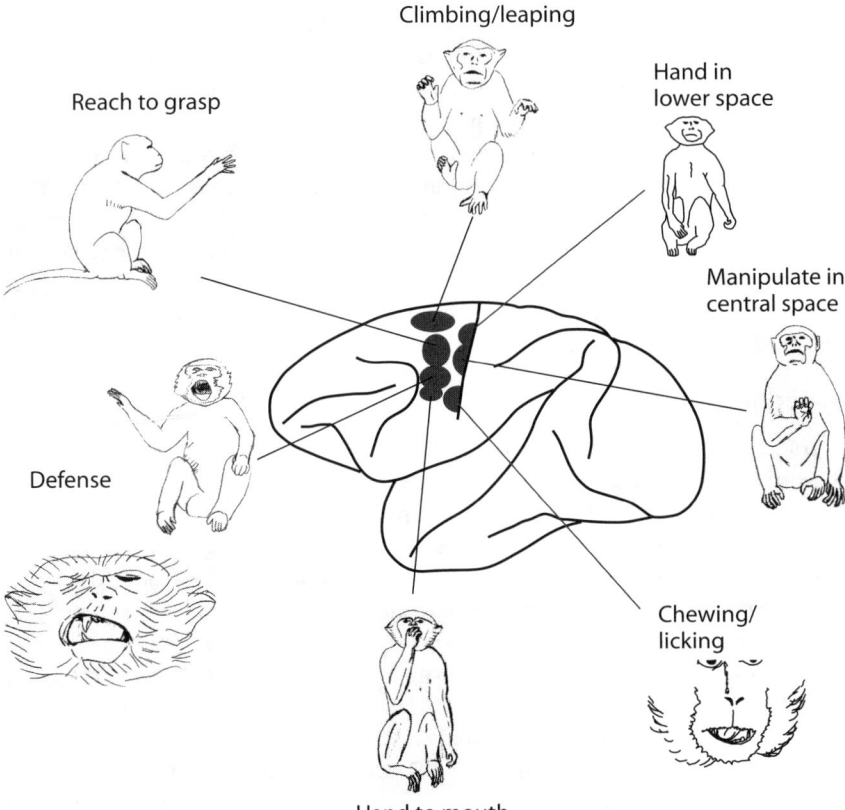

Figure 1-1 Action zones in the motor cortex of the monkey. Seven common categories of movement evoked by electrical stimulation of the cortex on the behaviorally relevant time scale of 0.5 sec. Images traced from video frames. Each image represents the final posture obtained at the end of the stimulation-evoked movement. Within each action zone, movements of a similar behavioral category were evoked. Based on results from Graziano et al. (2005; Graziano, Taylor, et al., 2002).

protect the face from an impending impact; and moving all four limbs as if leaping or climbing. The behavioral repertoire of the animal seemed to be rendered onto the cortical sheet. One might say that the cortical motor system had an action map.

The evoked movements were also roughly arranged across the cortex according to the location in space to which the movement was directed. The height of the hand was most clearly mapped across the cortical surface. Stimulation of the lower (ventral) regions of cortex commonly drove the hand into upper space, and stimulation of upper (dorsal) regions of cortex commonly drove the hand into lower space (Figure 1-2). Again, an important aspect of the animal's action repertoire was mapped across the cortex.

Over the next several years, as I set up my own lab at Princeton, we studied these cortical action maps with a variety of methods. We measured arm movement at high resolution to better understand the electrically evoked actions. We chemically activated or inhibited neurons at sites in the cortex and measured the effect on the monkey's behavior. We measured the neuronal activity in motor cortex that occurs during spontaneous movement to determine if the neurons are naturally tuned to complex actions. We even carried a video camera to the zoo, and then to an island populated by wild monkeys, to better understand the natural simian movement repertoire.

This line of experiments led us to propose two principles to explain the basic properties of the motor cortex. One principle concerned the topographic layout of the motor cortex, and the other concerned the neuronal mechanism by which motor cortex caused movement.

Topographic Organization

A traditional view of the motor cortex is that it contains a map of the body. This map was famously depicted by Penfield, whose homunculus diagram is shown in Figure 1-3. This traditional topographic scheme, however, does not capture the actual pattern of overlaps, fractures, re-representations, and multiple areas separated by fuzzy borders. The homunculus does not adequately describe the topographic organization. A current view of the motor cortex is that it can be divided into many distinct areas with separate functions (Figure 1-4). Yet the functions are largely not known, and the properties described thus far tend to vary across cortex in a graded fashion without hard borders. Rather than a set of separate areas, the pattern resembles a statistical distribution with clustering. Labeling those clusters with acronyms, drawing borders around them, and assigning functions to them may provide a convenient description but does not explain the principles behind the organization.

Based on our stimulation results, we proposed an underlying topographic principle for the motor cortex: the reduction of the many-dimensional space of the animal's movement repertoire onto the two-dimensional surface of the cortex. This reduction is similar to the problem in cartography of reducing the three-dimensional, curved globe onto a two-dimensional map, introducing unavoidable distortions and fracture lines. In the case of motor cortex, however,

1. Introduction

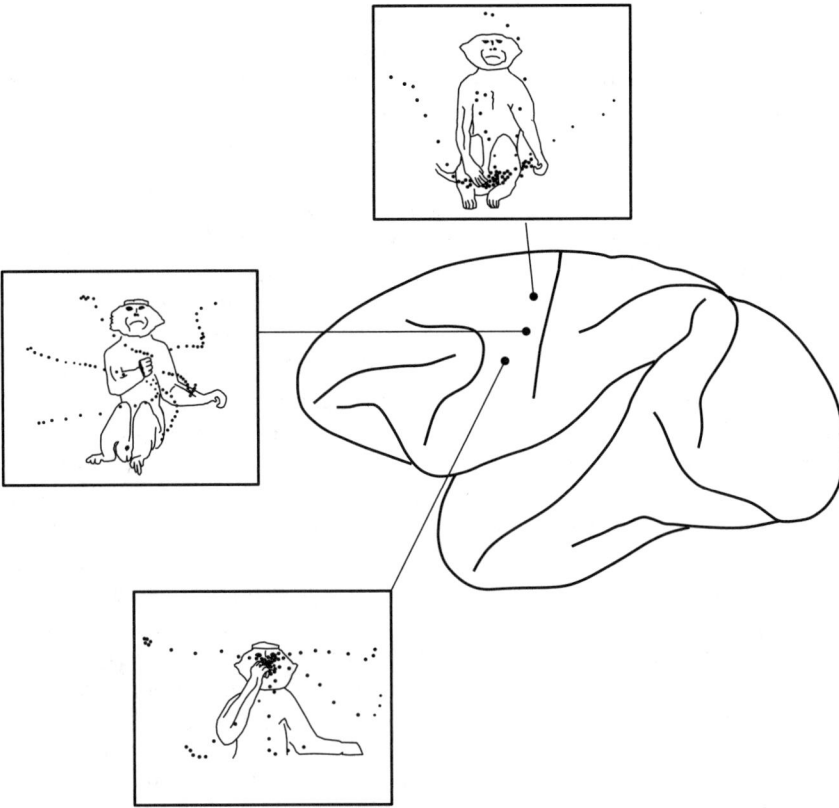

Figure 1-2 Progression of spatial locations to which hand movements are directed. Within the arm representation of the monkey motor cortex, electrical stimulation in dorsal cortex tended to drive the hand into lower space; stimulation in ventral cortex tended to drive the hand into upper space; stimulation in intermediate cortical locations tended to drive the hand to intermediate heights. Each image is a tracing of the final posture obtained at the end of a stimulation-evoked movement. Each dotted line shows the trajectory of the hand during the 0.5-sec stimulation train. Dots show the position of the hand in 30-ms increments. These trajectories show the convergence of the hand from disparate starting locations toward a final location. Adapted from Graziano, Taylor, et al. (2002).

the reduction is from the highly dimensional action space of the animal's normal behavioral repertoire to the two-dimensional cortical sheet. The core of this theory of cortical organization is that local continuity is preserved as much as possible. Information processors that need to interact are arranged physically near each other in cortex, presumably gaining a connectional advantage. One could term this principle of cortical organization the rule of "like attracts like." Perfect continuity is not possible, however, because of the unavoidable difficulties of rendering a highly dimensional space onto a two-dimensional sheet. The result is a complex compromise among many constraints.

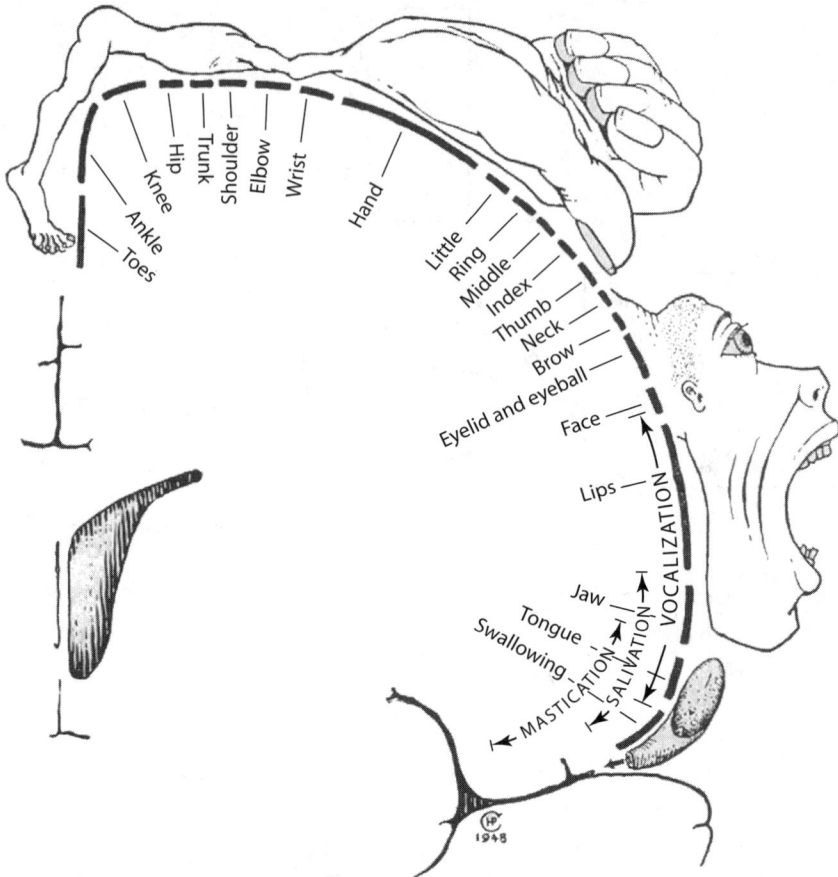

Figure 1-3 The motor homunculus of the human brain from Penfield and Rasmussen (1950). A coronal slice through the motor cortex is shown. Each point in motor cortex was electrically stimulated and the evoked muscle twitch was noted. Although each cortical point could activate many muscles, a rough body plan could be discerned.

In our proposal, the map of actions in Figure 1-1 is not by itself correct. It is present in the data, but the pattern is noisy and approximate. The map of hand locations shown in Figure 1-2 is also noisy and approximate, and therefore not by itself the correct description of motor cortex topography. The map of the body shown in Figure 1-3 is also present only in a rough sense and does not capture the complexities of the pattern. The proposal here is that all of these potential ways to organize movement, and perhaps others, are rendered onto the cortical sheet simultaneously, resulting in a compromise that does not neatly follow any single mapping dimension.

To test the validity of this theory of motor cortex organization, we used a mathematical model that collapsed an approximate description of the monkey's

1. Introduction

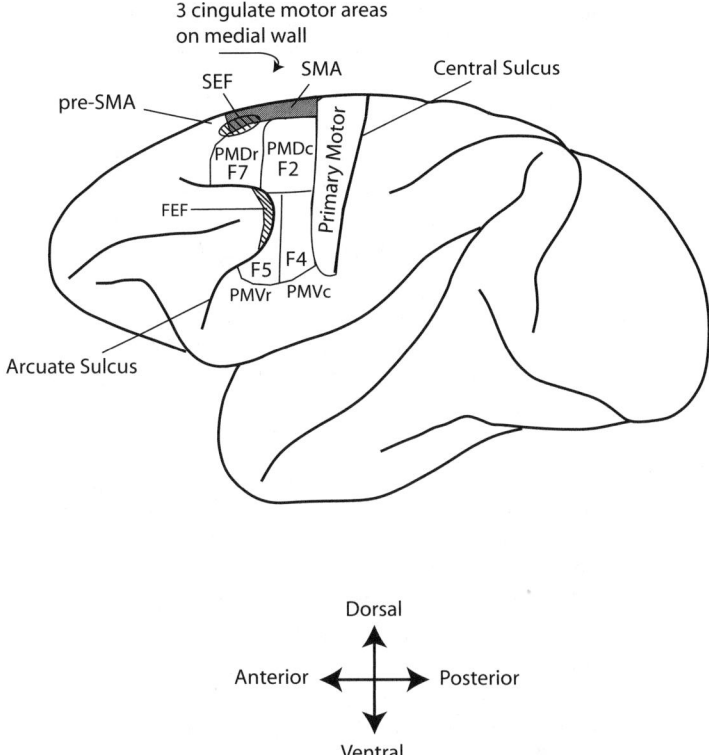

Figure 1-4 Some commonly accepted divisions of the cortical motor system of the monkey. PMDr = dorsal premotor cortex, rostral division, also sometimes called "Field 7" (F7). PMDc = Dorsal premotor cortex, caudal division, also sometimes called "Field 2" (F2). PMVr = Ventral premotor cortex, rostral division, also sometimes called "Field 5" (F5). PMVc = Ventral premotor cortex, caudal division, also sometimes called "Field 4" (F4). SMA = supplementary motor area. SEF = supplementary eye field, a part of SMA. Pre-SMA = pre-supplementary motor area. FEF = frontal eye field.

movement repertoire onto a two-dimensional sheet following the principle of maximizing local continuity (Aflalo and Graziano, 2006b; Graziano and Aflalo, 2007). The topographic organization generated by the model resembled the organization of the actual cortical motor system in many respects, including a rough clustering of movement categories as in Figure 1-1, an approximate mapping of hand position as in Figure 1-2, the outlines of a body map as in Figure 1-3, and the outlines of a primary motor area, dorsal and ventral premotor areas, supplementary motor area, frontal eye field, and supplementary eye field as in Figure 1-4. The theory of a dimensionality reduction was astonishingly successful in explaining the overarching organization of this large swath of cortex totaling about 20% of the macaque cortical mantle.

Mechanism of Movement Control

A traditional view of the neuronal machinery of movement control is that activity at a site in motor cortex propagates down a fixed pathway through the spinal cord, activating a set of muscles. Based on our stimulation results, however, the underlying mechanism seems to be less of a simple feed-forward pathway and more of a network. The effect of the network is to create a specific class of mapping from the cortex to the muscles, a mapping that can change continuously on the basis of feedback about the state of the periphery. If the periphery is relatively still, the mapping from cortex to muscles appears fixed and resembles the traditional view. But once the state of the periphery is allowed to vary as in natural movement, the mapping from cortex to muscles becomes somewhat fluid in a manner that facilitates complex movement control.

For example, when stimulation causes the hand to move to the mouth, different patterns of muscle activity are generated depending on the starting position of the limb. If the arm starts to the right of the mouth, stimulation evokes activity in the shoulder muscles appropriate for pulling the arm toward the left. If the arm starts to the left of the mouth, stimulation evokes muscle activity appropriate for pulling the arm toward the right. In effect, the mapping from the stimulated site in cortex to the muscles is not fixed. It changes depending on feedback information about the position of the limb. In this manner, the network can control limb position.

In general if the network receives feedback information about a specific variable, such as hand direction, or hand speed, or the posture of the arm, then the network can learn to control that variable. A network of this type is not limited to the control of one movement variable. It can in principle control muscle force directly and also control higher order variables, in combinations required for the performance of specific actions. A formal neural-network model that incorporated this principle of "feedback remapping" was able to control a model arm, successfully generating actions similar to those evoked in our stimulation experiments.

Theoretical Framework

The computational studies summarized above on topography and mechanism provide a potential theoretical framework for understanding at least the outlines of the motor cortex. In this framework, the purpose of the motor cortex is to control behaviorally useful actions in the motor repertoire; its complicated topographic organization is the result of a systematic rendering of the motor repertoire onto the cortical sheet; and the neuronal pathways between cortical neurons and muscles are designed to support the multijoint, feedback-dependant movements common in normal behavior. The goal of the present book is to elaborate on this theoretical framework.

1. Introduction

ORGANIZATION OF THE BOOK

The book is divided into two parts. The first part, ending in Chapter 6, reviews the previous literature from the discovery of motor cortex to the present, placing the current proposals into context. Any theory must be able to account for past results. By the same token, one cannot reject a theory because it fails to account for a distorted or mythological version of past results. One purpose of this review, therefore, is to lay to rest some of the common motor cortex myths, such as the myth of a muscle-by-muscle map. Chapter 6 discusses some advantages and limitations of the electrical stimulation technique because much of our work is based on this technique.

The second part, beginning with Chapter 7, describes the experiments and computational models that form the basis of the present perspective on motor cortex. Much of this work has been reported piecemeal in published articles. The present format allows for a more coherent global picture, additional analyses and results, and an extended discussion. Two chapters in particular are at the heart of the present story. Chapter 10 describes the proposal that the spatial layout of the cortical motor system can be understood as a reduction of the movement repertoire onto the cortical sheet. Chapter 11 describes the proposal that the mechanism of movement control by the motor cortex can be understood as a feedback-remapping mechanism, a divergent mapping from neurons in cortex to muscles that is continuously remapped based on information about the changing state of the periphery.

The final chapter of the book discusses possible links between motor control and social behavior, including the link between defensive movement and social smiles and between autism and abnormal movement control. The purpose of this final chapter is to emphasize the point that the motor system is not merely for activating muscles. It is a machine that allows intelligent interaction with the environment.

NOTE ON TERMINOLOGY

Figure 1-4 shows a schematic side view of a monkey brain with some commonly recognized cortical divisions (e.g., Dum and Strick, 2002; He et al., 1995; Luppino et al., 1991; Matelli et al., 1985; Matsuzaka et al., 1992; Preuss et al., 1996; Rizzolatti and Luppino, 2001). The cortical areas directly involved in motor control are typically divided into a lateral motor strip (unshaded in the figure) and a medial motor strip (shaded and partly hidden over the crown of the hemisphere). The lateral motor strip is divided into a posterior strip termed the "primary motor cortex," and an anterior strip termed the "lateral premotor cortex." The lateral premotor cortex is subdivided into a dorsal premotor area (PMD) and a ventral premotor area (PMV). In the monkey brain, each of these in turn is subdivided into a rostral area and a caudal area: PMDr, PMDc, PMVr, PMVc. These areas have also been labeled by Matelli et al. (1985, 1991) (in the same order) F7, F2, F5, F4. Because different groups have tended to publish work on different subdivisions, the PMDs are most often termed "PMDr" and "PMDc," whereas the PMVs are most often termed "F5" and "F4." A region with distinct properties that probably corresponds to the dorsal-most part of F4 has also been termed the "polysensory zone" (PZ) (Graziano and Gandhi, 2000). In the human brain, the divisions between rostral premotor and caudal premotor are less well established and the homology to the monkey brain is not yet clear.

The medial motor strip (shaded in the figure) was originally labeled the supplementary motor area, or SMA (Penfield and Welch, 1951). However, this region has now been subdivided into SMA, pre-SMA that lies directly anterior to SMA (Matsuzaka et al., 1992), and in the monkey a set of at least three little-studied areas on the medial part of the hemisphere buried in the cingulate sulcus, that are termed the "cingulate motor areas" (Dum and Strick, 1991).

Two gaze-control areas are also shown in cross-hatching in Figure 1-4. The frontal eye field (FEF) lies directly anterior to the arcuate sulcus and in its anterior bank. The supplementary eye field (SEF) lies within the anterior part of SMA. Both of these gaze areas are defined by the eye and head movements that can be evoked by electrical stimulation. Eye movements can also be evoked to a lesser extent from PMDr and PMDc (Bruce et al., 1985; Fujii et al., 2000).

The term *premotor cortex* is used to refer to at least three different regions. First, it has sometimes been used to refer to the lateral premotor cortex (PMDr, PMDc, PMVr, and PMVc). Second, it has been used to refer specifically to the dorsal part of the lateral premotor cortex (PMDr and PMDc). Third, it has been used to refer to all cortical motor areas excluding the primary motor cortex. The looseness with which the term is used can lead to some confusion.

The term *motor cortex* originally referred to the lateral motor strip, when that area was believed to be the only motor map of the body. It is now used variously to indicate the primary motor cortex, the lateral motor strip including primary motor and lateral premotor cortex, all cortical motor areas inclusively, or whatever part of the cortical motor system is under discussion at the moment. Because one theme in this book is that the divisions among motor areas are not as clear as sometimes suggested, it is useful to have a term that is intentionally ambiguous.

Chapter 2

Early Experiments on Motor Cortex

INTRODUCTION

This chapter describes how the dominant ideas about motor cortex first emerged. Many of the forgotten initial observations are still of direct scientific relevance. Moreover, the history shows how myths and factoids evolved and became resistant to change. Tracing these scientific stories reminds us that the prevailing beliefs at any time are not to be trusted. Certain beliefs, such as the early view that the cortex is inexcitable, or the more recent view of a discrete somatotopic map of the body in the primary motor cortex, are repeated and simplified through repetition until they become parables of uncertain validity.

This chapter traces motor cortex research from its beginning to the motor maps of Penfield and Boldrey (1937) and Woolsey et al. (1952). This segment of the history is mainly about electrical stimulation applied to the surface of the cortex. Using this technique, researchers drew motor maps of greater and greater elaboration. After Penfield and Woolsey, more fine-grained techniques such as microstimulation and single-neuron recording were used to probe the details and, as might have been expected, reopened all the same questions and debates. The more modern story of motor cortex, post-1952, is summarized in Chapters 3 through 5.

SWEDENBORG

There is some variation of opinion about where to begin the history of motor cortex research. Gross (1997) describes the remarkable case of Emanuel Swedenborg, a Swedish philosopher and mystic of the eighteenth century. In 1744 Swedenborg wrote a treatise on the brain. He proposed, among other remarkably accurate hypotheses, that movement was controlled by the cerebral cortex; that the feet were controlled by the uppermost part of the cortex; that the midsection of the body including the abdomen was controlled by the midregion of the cortex; and that the face was controlled by the lowermost part of the cortex. At that time the prevailing view of the cerebral cortex was of a nutritive or protective rind that served no mental function (Gross, 1997), yet Swedenborg correctly described the functional importance of the cortex and the upside-down topography of the motor map. Unfortunately his writings do not describe how he deduced these properties of the cortex. He is known to have visited contemporary physiology labs and may have observed a set of suggestive experiments that were never independently published. In any case,

Swedenborg appears to have been the first to propose a topographic motor map in the brain, predating Fritsch and Hitzig by 130 years. His views, however, were not generally known in his time and had little or no impact (Gross, 1997).

TODD

Reynolds (2004) begins his history of motor cortex with Robert Bentley Todd, an Irish physiologist who in 1849 published a set of observations and speculations on epilepsy. Todd (1849) attempted to define the brain regions that caused epilepsy. He argued that only the midbrain and cerebral cortex are likely to be involved. His evidence for midbrain involvement was that, when he electrically stimulated the midbrains of rabbits, he evoked general convulsions that resembled epilepsy.

On the cerebral cortex, Todd's observations were less direct. The prevailing view at that time was that no movement whatsoever could be evoked from the cortex by any stimulation. It was "inexcitable." Yet Todd noted that on postmortem examination, epileptics were sometimes found to have visible damage to the cerebral cortex. He described a case study of a two-year-old boy who developed seizures localized to his left hand. Over several days the seizures grew worse and spread to the entire left side of his body. On his death shortly after, the boy was found to have lesions of the dura over the right cerebral hemisphere. Todd deduced that an irritation or malfunction of the cerebral cortex on the right side resulted in muscular seizures of the left limbs.

Todd came remarkably close to deducing the motor functions of the cortex. In his final analysis, however, he shot very wide of the mark. His interpretation, in hindsight, seems to be an attempt to force his observations into the prevailing view that the cerebral cortex was inexcitable. Todd suggested a three-part mechanism: malfunction of the cerebral cortex was responsible solely for the loss of consciousness and higher mental functions during epilepsy; malfunction of the midbrain was responsible for the motoric convulsions; and malfunction of the spinal cord and medulla was uncommon and resulted in sustained muscle contractions rather than epileptic seizures. If the cortex played a role in movement control, it was only secondarily by way of the midbrain. In his words, "Under ordinary stimulation of the substance of the hemispheres, the fibres are incapable of exiting motion. It is not the office of these fibres to propagate the nervous force to muscles but to other nervous centres" (Todd, 1849, p. 999). Although correctly identifying a link between epilepsy and the cerebral cortex, Todd could not escape the views of his time and therefore missed the point of the cortical control of movement.

BROCA

One could almost start the history of motor cortex research with Paul Broca, the French neurologist who in 1861 described the case study of Tan. Although Broca met Tan only days before the patient's death, Broca was able to reconstruct some of Tan's past from interviewing hospital staff and friends. At the age of thirty one, Tan was admitted to the Hospice of Bicetre with an inability to speak. In all other respects he was normal and apparently intelligent, but other than the word "tan-tan" (from which he acquired his nick name) and an unrecorded gross swear word that he used when frustrated, he had lost the ability to produce vocal language. Over the next twenty-one years he slowly developed a weakness and then a paralysis of the right arm, followed by a similar paralysis of the right leg. He became bedridden. Because his sheets were changed only once a week, he developed an infection of the right leg that was not noticed until after it had become life threatening. It was at this point that Broca examined him. A few days later the patient died, and Broca performed an autopsy of the brain. The left frontal lobe was badly degenerated with a focus of degeneration in the third frontal convolution, now commonly known as Broca's area.

Broca concluded that the lesion must have begun small, affecting the specific cortical center for speech, and then gradually spread to surrounding tissue, including the precentral gyrus (now known to be the site of motor cortex). This spread of the degeneration, according to Broca, must have caused the gradual paralysis of the body. Broca, like Todd, came within a hair's breadth of deducing the motor map. His detailed observations placed him in exactly the right part of cortex. His careful estimates of the center of the degeneration and its rate of spread could have led him to a cortical map arranged sequentially from one body part to the next, from face to arm to leg, in the order of the progression of Tan's symptoms. He was willing to infer that at least one function, speech, was localized to a region of the cerebral cortex. Yet in his words, "Everybody knows that the cerebral convolutions are not motor organs. The corpus striatum of the left hemisphere is of all the attacked organs the only one where one could look for the cause of the paralysis" (Broca, 1861/1960, p. 70). Because of his acceptance of the beliefs of the time, he was totally unable to see the importance of his observations for motor control. Surely the lesson here is to be most wary of the thing that "everybody knows."

JACKSON

Most historical reviews of motor cortex research begin with the English neurologist John Houghlings Jackson (e.g., Ferrier, 1873; Foerster, 1936; Hitzig, 1900; Penfield and Boldrey, 1937). Jackson is generally credited with having deduced the existence of a somatotopic motor map in the cortex on the basis of the spread of epileptic seizures across the body. A close reading of his work, however, shows that this common belief about Jackson is completely wrong. It is an interesting case study in the way that historical myths become established in science. Jackson is something of the Nostradamus of neuroscience; his writing is ambiguous and rich enough that one can read almost anything into it. For the sake of getting the story right, it seems worth detailing the frankly brilliant ideas that Jackson actually did propose.

During the 1860s Jackson studied a large number of epileptic cases that he summarized in a publication in 1870. Some of his patients suffered from global seizures that simultaneously affected the entire body. Others suffered from partial seizures that began in one location on one side of the body. Jackson focused his theoretical work on the cases of partial seizures. In his writing he was quite clear that the seizures were caused by malfunction of the cerebral hemispheres, but he was inconsistent on whether they were caused by malfunction of the cortex or of the striatum, a large nucleus underlying the cortex. Jackson implied that the partial seizures, being simpler, were probably caused by the striatum, whereas the global seizures, being more complex, may have been caused by the cortex. In this respect Jackson fell into the same trap as Todd (1849) and Broca (1861/1960). In every case of a partial seizure in which he was able to examine the brain afterward, Jackson described damage to the cerebral cortex, not to the striatum; yet he seemed unable to let go of the idea that the cortex was too complex a structure for the control of a body part.

Jackson (1870) noticed that the partial seizures almost always began in the hand, in the face around the mouth, or in the foot, the three parts of the body that are most commonly used, or in his description, that have the most "varied uses." Furthermore, "fits which begin in the hand begin usually in the index finger and thumb; fits which begin in the foot begin usually in the great toe" (p.10). From these observations he deduced that the amount of neurons in the brain devoted to a body part, and therefore the chances that the body part may be affected by seizures, must be proportional to the amount of use of the body part. Jackson therefore brilliantly formulated the general principle of brain organization that behaviorally important functions have physically larger representations.

Jackson also noticed that seizures beginning on the right side of the face and tongue were often followed by a lingering loss of speech. He inferred that these right-sided facial seizures were caused by nervous instability and explosive discharge in Broca's recently described speech area in the left hemisphere (Broca, 1861/1960). After the discharge, the brain area must suffer from fatigue resulting in a loss of speech. Jackson further deduced that because seizures beginning in the hand and the foot did not affect speech, they must be caused by instability of other, separate brain regions. He therefore correctly deduced that the brain contained different centers for the control of different body parts.

Jackson (1870) noticed that, "When a fit begins in the hand it goes *up* the arm and *down* the leg ... Now patients who have fits beginning in the foot tell me that the spasm goes *up* the leg and *down* the arm" (p. 23). This progression of convulsions from one body part to the next is now known as a Jacksonian march. This observation was not news to the patients who commonly tried to block the march by tying ligatures around their limbs. One woman with a seizure that always began in the hand would tie a ligature around her wrist; and when that did not work, she tied it higher up the arm.

2. Early Experiments on Motor Cortex

Another patient with seizures that began in the foot similarly tied a ligature around his ankle.

To Jackson, this spread of epilepsy from one body part to the next presented a theoretical problem. He believed that a partial seizure was caused by abnormal instability in a focused spot in the brain. In his words, "The fact that the symptoms are local implies, I hold, that there *is* of necessity a *local* lesion" (Jackson, 1870, p. 24). Yet a seizure that starts in the hand may spread to other body parts and in some cases may spread to the entire body bilaterally. Therefore, to Jackson, all the parts of the body affected by the seizure must be represented within the local, diseased brain region. He asks, "Why, if face, arm and leg are represented together in the square inch, is the fit a sequence only? Why are not all these parts convulsed contemporaneously?" (p. 27). His answer is that the diseased "square inch" of the brain must normally control useful sequences of actions. During a diseased discharge, the sequence of spasms or the spread across the body is a crude caricature of the stored sequence of actions.

Even by 1875, five years after Fritsch and Hitzig published their physiological map of motor cortex, Jackson (1875) wrote unambiguously: "When we grasp ... the more strongly the hand is used, the farther up the arm does the movement spread" and therefore, "if a fit begins in the thumb and index finger, there will probably be developed ... that series of movements which in health serves subordinately when the thumb and index finger are used" (p. 69). Jackson, therefore, failed totally to appreciate the true reason for the progression of partial seizures across the body, namely the spread of an epileptic storm across a map of the body in the cortex. He accepted the lay belief that tying a ligature around the limb can stop the spread of epilepsy up the limb, a view that is totally untenable if the spread is actually across the cortex. Jackson had no concept of the spread of epilepsy across the cortex, nor had he any concept of a somatotopic map in the cortex.

The mechanism of motor control that Jackson deduced was a collection of centers, possibly in the striatum, each one of which controlled the entire body with an emphasis on coordinating the action of one particular body part. Centers emphasizing the hand, face, and foot were larger than centers emphasizing other body parts because the hand, face, and foot required a more complex and varied movement repertoire. In these deductions he came remarkably close to the truth, but not as close as is sometimes suggested.

Jackson saw many of his speculations confirmed in 1870, when Fritsch and Hitzig stimulated the dog brain and demonstrated a set of distinct centers that corresponded to different parts of the dog's body. These centers were located in the cerebral cortex rather than in the striatum; but aside from his localization error, Jackson's essential concepts appeared to have been vindicated.

As physiological evidence for his movement centers accumulated, Jackson developed an overarching description of the brain basis of behavior. In 1890 he proposed that the control of movement could be divided into three levels that corresponded to three stages in animal evolution (based on a flawed understanding of Darwin's theory of evolution typical of that time, in which evolution progressed from lower to higher levels). The lowest level of motor control was subcortical, controlling the simplest elements of movement. The middle level lay in the motor cortex of Fritsch and Hitzig, controlling movement in an integrated and complex fashion. The highest level lay in the prefrontal cortex, controlling movement in the most abstract sense of will or thought. Epileptic seizures of the highest level caused a loss of consciousness; seizures at the middle level caused muscular convulsions; seizures at the lowest level normally did not occur and artificial stimulation here caused sustained contractions. It is only fair to point out that this influential idea indelibly associated with Jackson's name is not entirely original to Jackson. It is an elaboration and modification of Todd's 1849 proposal. Although Todd did not express the idea as clearly and attributed his functions to a slightly different underlying anatomy, he outlined a highest level of consciousness that lay in the cortex (seizures here induced loss of consciousness), a middle level of movement control that lay in the midbrain (seizures here induced muscle convulsions), and a lowest level of movement control that lay in the spinal cord and medulla (normally seizures did not occur here and direct stimulation caused sustained muscle contraction).

Jackson was particularly clear that none of the three levels of movement control, not even the lowest level, was a simple activator of muscles. He wrote: "motor centers of every level represent movements of muscles, not muscles in their individual character" (Jackson, 1890, p. 687). By 1890, therefore, the "muscles vs. movements" debate of motor control was fully framed.

DISCOVERY OF MOTOR CORTEX: FRITSCH, HITZIG, AND FERRIER

In 1870, the two German scientists Fritsch and Hitzig published a set of experiments on the cortex of the dog brain. Borrowing Frau Fritsch's dressing table as an operating surface, they exposed the brains of dogs (unanesthetized and anesthetized; the results were similar) and stimulated the cortex using brief electric discharges from a battery. In the anterior region of cortex, this stimulation evoked muscle twitches. Fritsch and Hitzig described distinct cortical centers from which twitches of different body parts could be obtained. Figure 2-1 shows a redrawing of their summary map of cortical centers. They described five centers, indicated in the figure by the five cortical sites at which optimal results could be obtained. For example, stimulation at one site evoked movements of the foreleg; stimulation near the site evoked these movements less well; stimulation of sites between the labeled points evoked either no movement or, if the electrical current was turned up, combined movements of the adjacent centers. In two animals, Fritsch and Hitzig removed part of the foreleg center on the right side and observed that the dogs were thereafter unable to move the left foreleg in a coordinated fashion while walking or running.

Fritsch and Hitzig may have been steeped in the beliefs of the time, but unlike previous investigators of movement control, they took their data at face

2. Early Experiments on Motor Cortex

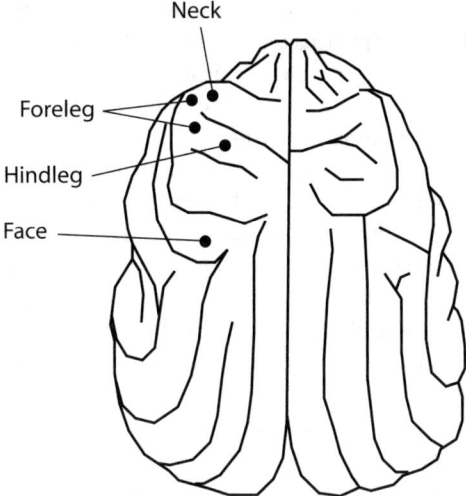

Figure 2-1 The map of stimulation-evoked movements in a dog brain adapted from Fritsch and Hitzig (1870/1960). Each point indicates the approximate location of a movement center. Stimulation at or near each point evoked movements of the indicated body part.

value. They drew two broad conclusions. First, contrary to the entrenched belief, the cerebral cortex was in fact "excitable" in that stimulation of it caused overt behavior. Second, contrary to Flourens' notion (1824/1960) of a homogenous cerebral cortex (at least in a chicken, Flourens' animal of choice), in which all functions were equally distributed, Fritsch and Hitzig (1870/1960) found that the front or anterior half of the dog cortex "stands in immediate connection to muscular movements" while the back or posterior half "has evidently nothing to do with it" (p. 92).

Note that Fritsch and Hitzig did not describe a somatotopic map of the body in the modern sense. Beginning students of neuroscience today are taught that there is an elaborate roster of body parts separated and placed in order along the cortex. The map of Fritsch and Hitzig, however, is not so precise. First, the "map" is not continuous but instead is divided into five islands each one surrounded by relatively inexcitable cortex. Second, no clear topography is reported within each island; instead each island represents an undifferentiated collection of many muscles. Third, the rough overall arrangement is not topologically correct. The neck is represented at the anterior end of the "map" and the face is represented at the posterior end behind the hind leg.

Fritsch and Hitzig came down clearly on the side of the cortical control of movements rather than muscles. In their view, a cortical center was "a middleman … in which a similar but better coordination of muscle movement takes place than in the gray substance of the spinal cord or brain stem" (Fritsch and Hitzig, 1870/1960, p. 92). Furthermore, they understood that movement control cannot be separated from sensory processing. After lesions of the foreleg center, the dogs not only lost the ability to control coordinated movements of the leg but also "had obviously only an imperfect consciousness of the shape of their member, they had lost the faculty to make a complete conception of it" (Fritsch and Hitzig, 1870/1960, p. 96). This view of highly complex cortical centers that controlled movement by combining sensory and motor processing closely resembled Jackson's concept of motor centers, although it appears that Fritch and Hitzig were unaware of Jackson's work at the time.

Shortly after Fritsch and Hitzig's initial discovery of motor cortex in the dog brain, Ferrier (1873) replicated the results and extended them from dogs to cats, rabbits, and guinea pigs, though he was unable to evoke movements from the brains of birds, complaining that they were too soft and oozed when stimulated. He went on to study monkeys (Ferrier, 1874), establishing the motor map on the precentral gyrus of the primate brain with the leg represented in a dorsal location and the mouth represented in a ventral location. Figure 2-2 shows a redrawing of Ferrier's motor map with seven cortical regions arranged along the precentral gyrus, numbered as he originally numbered them. In addition to mapping the motor cortex along the precentral gyrus, Ferrier also evoked eye and head movements from cortical regions that

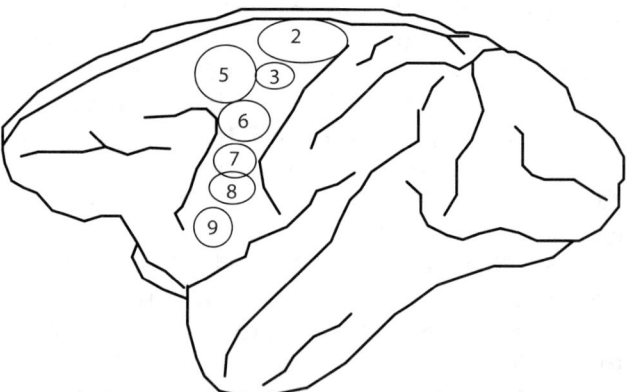

Figure 2-2 The map of stimulation-evoked movements in a monkey brain adapted from Ferrier (1874). The cortical regions arranged along the precentral gyrus are numbered as he originally numbered them. The effects of stimulation were: Region 2, leg and foot movement. Region 3, trunk, leg, and tail movement. Region 5, combined arm and leg movement. Region 6, arm and hand movement, including some hand-to-mouth movements. Regions 7–9, mouth movement.

2. Early Experiments on Motor Cortex

became known as the frontal eye field (FEF) and the posterior eye field (now usually known as the lateral intraparietal area or LIP, though Ferrier was not as specific about its location).

Ferrier's experiments differed from Fritsch and Hitzig's in several important respects. First, one must credit Ferrier with the first true description of a motor map in the modern sense of map. He described the somatotopic progression in detail.

Second, rather than use a direct current pulse from a battery as Fritsch and Hitzig had done, Ferrier used an alternating current that could be extended over several seconds. By extending the stimulation, he found that the previously described muscle twitches unfolded into longer, apparently coordinated actions that he called purposive movements. "The movements ... resulting from excitation of the individual centers are purposive or expressional in character, and as such we should, from psychological analysis, attribute to ideation and volition if we saw them performed by others" (Ferrier, 1873, p. 73). For example, in one experiment, stimulation within the center for control of the leg (circle 2 in Figure 2-2) caused an action "just such as when a monkey scratches its abdomen with its hind leg" (Ferrier, 1874, p. 413). In another experiment, stimulation within the center for the hand and arm (circle 6 in Figure 2-2) "brings the hand up to the mouth, and at the same time the angle of the mouth is retracted and elevated" in a manner resembling a feeding movement (Ferrier, 1874, p. 418).

Ferrier argued with Fritsch and Hitzig over the correct method of stimulation. In Ferrier's view (1873, 1874), the shorter stimulation missed essential movements, whereas in Hitzig's view (1900) the longer stimulation evoked seizures and therefore produced artifactual results. Despite the squabble, their views on motor cortex were almost identical. They both believed that motor cortex contained a set of cortical centers that controlled movement at a high level, coordinating groups of muscles in a meaningful fashion rather than controlling individual muscles.

A lecture that Hitzig published in English in 1900, thirty years after the initial discovery of motor cortex, is a vivid editorial on his contemporaries and reveals something of the issues surrounding the research at that time. He is mainly complimentary of Jackson of whom he predicts, "His thoughts will ever again rise from the seemingly lifeless dust, and will spur posterity on to renewed intellectual labour in the field that he cultivated" (p. 546). He states that Goltz had "given occasion to unpleasant conflicts and thereby also to the spreading of great confusion over the questions with which we are busied" (p. 558). Goltz had suggested that Fritsch and Hitzig's (1870/1960) findings were caused by the spread of electric current from the surface of the cortex to the underlying striatum. Hitzig argues forcefully against Ferrier's view that the motor cortex is dedicated to movement control and does not participate in sensation. He argues equally forcefully against Schiff's view that the motor cortex is purely a sensory area with no true motor function. He systematically antagonizes every one of his contemporaries and then ends the address with the uplifting platitude that scientists form "an army that knows no separation into different camps ... in the battle against ignorance" (Hitzig, 1900, p. 581).

ELABORATION OF THE MOTOR MAP: BEEVOR AND HORSLEY

Beevor and Horsley (1887, 1890) conducted a series of experiments that greatly extended the findings on motor cortex but also, in some ways, retreated from the initial insights of Fritsch and Hitzig (1870/1960) and of Ferrier (1874).

In 1887 Beevor and Horsley published a study focusing on the arm and hand representation in the monkey motor cortex. The study built on Ferrier's (1874) work in that it examined the complex or "purposive" movements evoked from the cortex. The authors outlined a cortical map that was not merely a plan of the body's muscles but was an arrangement of useful movements. I find this study particularly interesting because it predicts many of our own observations one-hundred-and-ten years later (Graziano et al., 2005; Graziano, Taylor, et al., 2002). Beevor and Horsley's study may be the last to examine the topographic mapping of complex actions on the motor cortex until ours.

In their experiment, Beevor and Horsley found that stimulation of the lower or ventral part of the arm representation tended to evoke movements consistent with bringing an object toward the body, especially toward the mouth for feeding. These movements included flexion of the elbow, supination of the forearm (orienting the palm toward the body), and closing of the grip. These authors therefore roughly located the hand-to-mouth zone that was redescribed much later (Graziano, Taylor, et al., 2002). Because the hand-to-mouth zone can be several millimeters wide, it is reasonable that Beevor and Horsley were able to localize it using their relatively coarse surface stimulation through two electrical poles spaced 2 millimeters apart.

Beevor and Horsley also found that stimulation of the upper or dorsal part of the arm representation tended to evoke "advancing" movements of the arm including a pronation of the forearm and opening of the hand, which they variously interpreted as "defensive" or, in a later paper that extended the same work (Beevor, 1888), as "reaching forward ... to seize" an object. They described a specific, circumscribed region of the precentral gyrus that, when stimulated, consistently evoked this open-hand arm-extended movement. This region is a close match to our reach-to-grasp zone from which we evoked precisely the same movement (Graziano et al., 2005). It is also a close match to the modern caudal dorsal premotor cortex (PMDc) thought to play a role in reaching.

In addition to studying the complex actions evoked at each cortical site, Beevor and Horsley also studied what they termed the "primary" movement, the initial or at least the most visible twitch to occur with a short pulse of stimulation. They noted that the cortical map of primary movements followed the approximate plan of the body, whereas the cortical map of purposive actions, or of the primary, secondary, and tertiary movements evoked with

2. Early Experiments on Motor Cortex

longer stimulation, was much more overlapping. In their discussion they imply that the two types of map, a somatotopic map of the body and a map of complex movements, do not negate each other, and that both are useful ways to view the cortical organization.

In their later experiments, however, Beevor and Horsley confined their investigations to the primary movements, perhaps because they were easier to study or presented fewer interpretational complexities. In 1890 they published a meticulous study in which they dissected away the cortex overlying the motor area and stimulated the severed ends of fibers. In this manner they showed that the movements evoked from the motor cortex depended on the fibers of the pyramidal tract that originate in cortex and course down through the internal capsule, the midbrain, and the medulla to the spinal cord. On stimulating these fibers they obtained a detailed map of the body that precisely matched the map in the overlying cortex. Their map of motor cortex is redrawn in Figure 2-3.

Beevor and Horsley's 1890 experiment represented a move toward a "muscles" view of motor cortex function and away from a "movements" view. This study suggested that the movements evoked by stimulating the cortex were caused by the cables that ran down to the spinal cord; the motor cortex itself was merely the start point of these cables; and the intrinsic processing in the cortex was of minimal importance because the same map was obtained whether one briefly stimulated the cortex or the cables. Beevor and Horsley's map (Figure 2-3) was a detailed stacking of body parts. In some parts of the map a single joint such as the shoulder, elbow, or wrist was assigned a discrete area. This view of motor cortex was radically different from the view presented in their earlier (Beevor and Horsley, 1887) paper or proposed by Ferrier (1874)

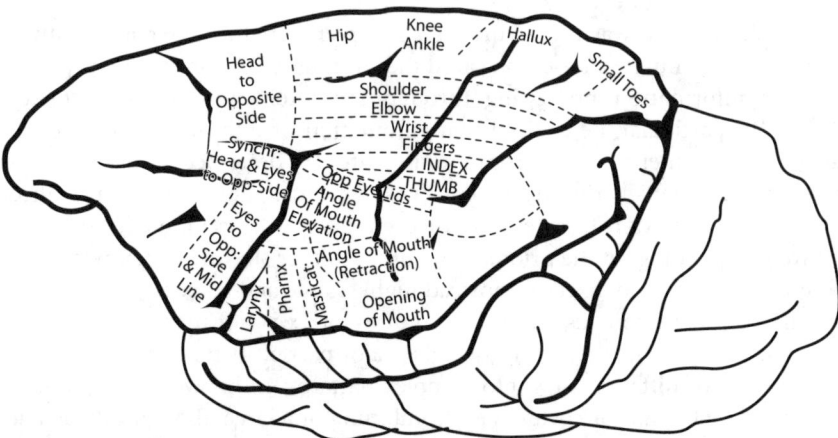

Figure 2-3 The map of stimulation-evoked movements in the monkey brain from Beevor and Horsley (1890).

or Fritsch and Hitzig (1870/1960). In the earlier view, motor cortex was a collection of motor centers. Each center was a sophisticated processing device that produced meaningful actions by intelligently combining the movements of many muscles and joints and perhaps also by integrating sensory input with motor output. Beevor and Horsley's 1890 paper instead implied a view in which there were no motor centers, and in which the function of motor cortex was defined by its descending connections to the spinal cord—an antinetwork, feed-forward view that, unfortunately, came to dominate the twentieth-century research on motor cortex and to some extent is still dominant today.

ELABORATION OF THE MOTOR MAP: SHERRINGTON

In 1901 Sherrington and his students began a series of unusually varied and revealing experiments on the motor cortex of anesthetized apes (e.g., Grunbaum and Sherrington, 1901, 1903; summarized in Sherrington, 1939). They electrically stimulated the motor cortex of twenty-two chimpanzees, three gorillas, and three orangutans.

To obtain as discrete a map as possible, Sherrington used surface stimulation that was as weak and brief as possible while still evoking movements. In this way Sherrington continued a trend away from studying the representations of complex actions. His map of a chimpanzee's motor cortex is shown in Figure 2-4. He suggested that this body plan on the cortex closely follows the plan of the spinal roots that innervate the muscles. He also suggested that the map is more discrete in "higher types" or apes, less discrete in monkeys, and even less so in dogs, because apes have the greater need to combine separate movements into novel combinations. (Sherrington, like most of his contemporaries and like Jackson before him, had not properly grasped Darwin's concept of evolution and therefore wrote about lower and higher animals as well as lower and higher levels of evolution.)

Yet Sherrington was too accurate an observer to reduce the cortex entirely into a map of muscles. He emphasized the complexity of cortex and its potential for information processing through the interaction between cortical points. In particular, he noticed that stimulation of one site in cortex could alter the movement evoked from another site. He described several types of interaction in which the movement evoked from a site could be modified, changed from flexion to extension, or even changed to a different body part entirely, depending on the prior stimulation of other sites. The map was therefore not anatomically fixed but instead highly labile and subject to experience. He arrived at the conclusion that the motor cortex is not merely a device for the descending control of the muscles; a major part of its function was the lateral linking of different sites into complex patterns. In his words, "a property possessed by the motor cortex is the combining of a large, though exhaustible, number of movements ... into sequences of very great variety" (Sherrington, 1939, p. 424). This view is similar to a view more recently suggested by Huntley and Jones (1991) and Schneider et al. (2002).

2. Early Experiments on Motor Cortex

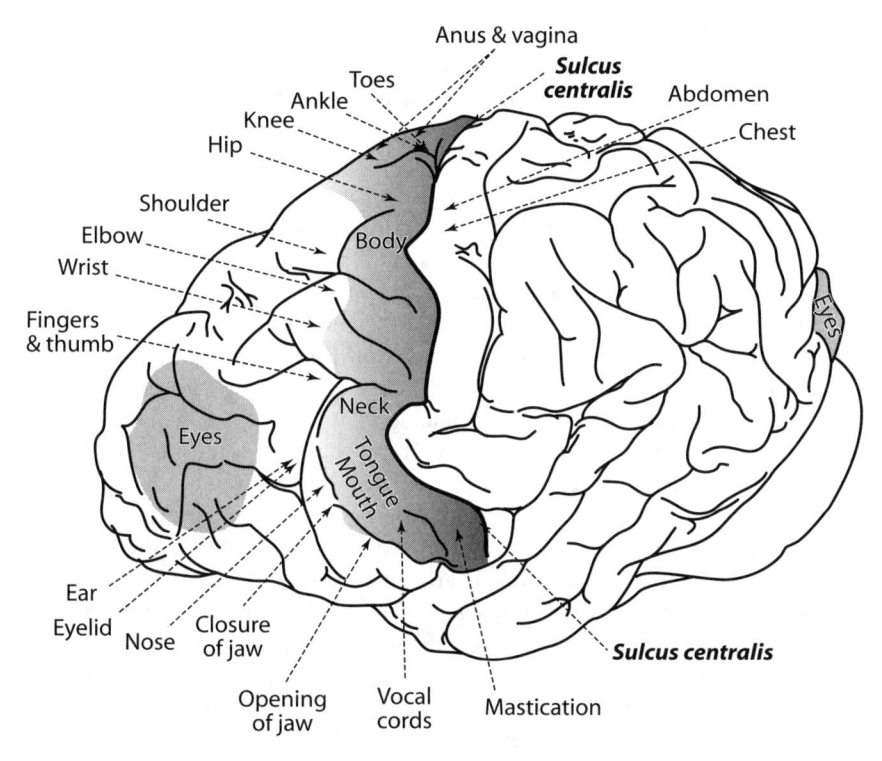

Figure 2-4 The map of stimulation-evoked movements in a chimpanzee brain from Grunbaum and Sherrington (1901). The stippled strip indicates the motor cortex. The stippled circular area represents the frontal eye field.

Sherrington made several other important observations on the motor cortex of apes. He noted that the posterior border of the excitable zone, the zone from which movements could be evoked, was sharply defined; but the anterior border on the precentral gyrus was more of a gradient than an edge. The anterior border was not only gradual but also unstable. If he stimulated at the posterior edge of motor cortex and marched the electrode forward in a series of stimulations until no more movement was evoked, he obtained a relatively anterior border and a relatively wide expanse of motor cortex. If he reversed the series, stimulating first in an anterior region outside of motor cortex and marching the electrode backward until a movement was evoked, he obtained a different border, a more posterior one that resulted in a narrow expanse of motor cortex. This observation is important given the later attempts to divide the precentral gyrus into a posterior, primary motor strip and an anterior, premotor strip. The physiological properties, at least as seen at this time, did not suggest a clear border between two different cortical motor areas. The differences took the form of a labile gradation, not a categorical distinction.

Finally, Sherrington (1939) performed a set of unusually revealing lesion experiments. After removing a region of motor cortex, he observed the specific loss of motor function and how the loss changed over the subsequent days. For example, after he removed the hand representation on the left side of a chimp brain, the chimp's right hand was weak and no longer coordinated. The chimp seemed to know what it wanted to do but couldn't perform the movements. In Sherrington's interpretation, "Surprise at the failure of the limb to execute what it intended seemed to be the animal's mental attitude" (p. 435). Sherrington therefore suggested that the lesion must have caused a "defect in the motor execution rather than in the mental execution of the act" (p. 435). The intention to make a movement, in his speculation, was not contained within motor cortex; it must be localized somewhere else in the brain, perhaps in the parietal lobe. This insightful speculation foreshadows the much more recent work of Andersen and colleagues (e.g., Snyder et al., 1997) on movement intention represented in the parietal lobe.

Within a month after the lesion, the chimp had recovered. It was able to move its right hand in a coordinated fashion. Some other brain area must have taken over the control of the hand. In a series of secondary lesions, Sherrington was not able to find the cortical region responsible for the regained function. Lesioning the cortex directly around the original lesion, the more distant cortex representing other body parts, or the hand representation in the opposite hemisphere, did not take away the regained function in the right hand. This result suggested that the motor cortex as understood at the time must not be the only area of cortex capable of controlling voluntary movement. In effect, Sherrington had provided evidence that the cortex probably contained more than one motor area, and that the others were yet to be discovered.

CAMPBELL AND THE PROPOSAL OF TWO DISTINCT MOTOR MAPS

Originally the term *motor cortex* referred to the motor map of the body along the precentral gyrus. There was no other motor map and no need for a more precise term. In 1905, Campbell published a division of cortex on the basis of the appearance of cells and fiber tracts. He suggested that the motor cortex was divided into two cytoarchitectonic regions: a posterior region that was characterized by a dense population of giant pyramidal-shaped cells (termed "Betz cells") in the deeper layers, and an anterior region that was similar in most respects but lacked the Betz cells. Campbell acknowledged that the two areas do not have an abrupt border. They grade into each other. Nonetheless he argued for a distinction between them. (One might say that cytoarchitectonics is the taking advantage of the human perceptual tendency to see borders where a gradient actually exists.) The posterior strip he termed the "precentral" cortex and the anterior strip he termed the "intermediate precentral" cortex (Figure 2-5). Brodmann (1909) saw a similar division of the motor cortex into two areas that he termed "area 4" (Campbell's precentral cortex) and "area 6" (roughly matching Campbell's intermediate precentral cortex). Today Brodmann's terminology is more common.

2. Early Experiments on Motor Cortex   27

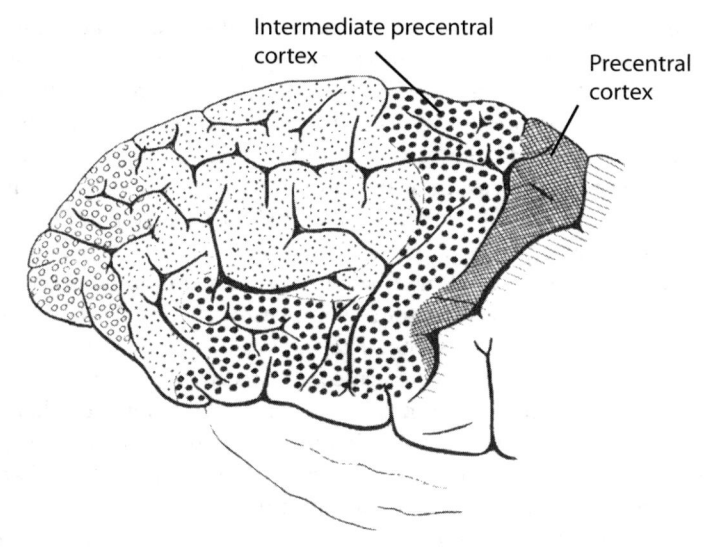

Figure 2-5 The division of the human motor cortex into a posterior strip, the "precentral" cortex, and an anterior strip, the "intermediate precentral" cortex, from Campbell (1905).

Campbell invoked the three motor levels of Jackson (1890). The lowest level was in the spinal cord and medulla. The middle level, according to Campbell's proposal, was the precentral strip characterized by Betz cells. Campbell viewed this region of cortex as the true or "primary" motor cortex in direct control of muscles. In this way Campbell officialized the term *primary motor cortex*, assigning Beevor and Horsely's (1887) "primary" movements to a specific sector of cortex. Finally, Campbell placed the highest level of motor control, involved in the coordination of complex actions, at least partly in his proposed intermediate precentral cortex.

Campbell amassed a set of reasons to support this division of motor cortex into two motor areas, a "primary" one and an "intermediate precentral" one. Campbell first argued that the giant Betz cells were the critical output cables from the cortex controlling movement. This speculation turned out to be incorrect. The Betz cells compose only about 3% of the descending projection from the cortex to the spinal cord in primates (e.g., Lassek, 1941). Yet to Campbell the Betz cells were the primary conduit by which the cortex controlled the body. He examined the brains of patients who had suffered progressive loss of muscle control, and he detected no abnormality except a loss of giant Betz cells in motor cortex. He studied the brains of amputees and discovered a loss of Betz cells in the region of motor cortex corresponding to the missing limb, as if, lacking muscles to control, the Betz cells had died. Indeed Campbell rather brilliantly demonstrated the map of the body in the human motor cortex by examining the patterns of degeneration caused by different amputations.

Because in Campbell's (incorrect) view the Betz cells formed the main or only motor output from cortex, and because according to Campbell the Betz cells were found essentially in the posterior motor strip and not the anterior strip (a view that was also not quite correct; the distinction is more a gradient than a border; e.g., Bucy, 1935), therefore only the posterior strip was the true or primary motor area whereas the anterior strip must serve a different function.

Campbell's arguments for the complex functions of the anterior strip, the intermediate precentral cortex, were equally speculative. Broca's area was clearly responsible for the high-level organizing of movements into speech because lesions to it eliminated speech without eliminating the raw ability to move the mouth muscles (Broca, 1861/1960). The intermediate precentral cortex contained Broca's area, therefore the intermediate precentral cortex must generally control the complex coordination of movement. Note, however, that Campbell's intermediate precentral cortex (Figure 2-5) looks almost as though it has been gerrymandered to include Broca's area. Its ventral part extends like a foot with the toes in Broca's third frontal convolution.

Agraphia, a hypothetical syndrome in which a patient is selectively unable to write, Campbell speculatively localized to a region of the intermediate precentral cortex just anterior to the primary motor hand area, a region now thought to be mainly an eye movement area unrelated to hand movement. Campbell also speculated that a high-order leg area, controlling complex actions of the leg, must be located just anterior to the primary motor leg area, in a region now typically considered to be the supplementary motor cortex. The specifics of Campbell's hypothesis are therefore almost all wrong.

Campbell speculated beyond his data and arrived at a sequence of errors. Perhaps his fundamental error was an overreliance on the dubious doctrine that the function of a brain area can be deduced from its appearance under a microscope. Yet his two main suggestions were influential. He suggested first that the motor cortex could be divided into an anterior and a posterior area; and second, that the areas were hierarchically linked, the anterior one controlling the posterior one, which in turn controlled the spinal cord. These suggestions continue to resonate today.

VOGT AND VOGT AND THE ELABORATION OF THE MOTOR HIERARCHY

After Campbell, a range of experiments supported the hypothesis that the precentral gyrus was not uniform. For example, the German team Cecile Vogt and Oskar Vogt (1919, 1926) divided the monkey motor cortex into a large number of cortical fields on the basis of appearance under a microscope. Their divisions included a primary motor field and several ventral and dorsal premotor fields. Most of their physiological work focused on the dorsal motor areas, shown in Figure 2-6. Their first cortical field, area 4, was the most posterior region and was dense with Betz cells. They referred to this area as the primary motor area because stimulation at low currents evoked simple movement twitches of

2. Early Experiments on Motor Cortex

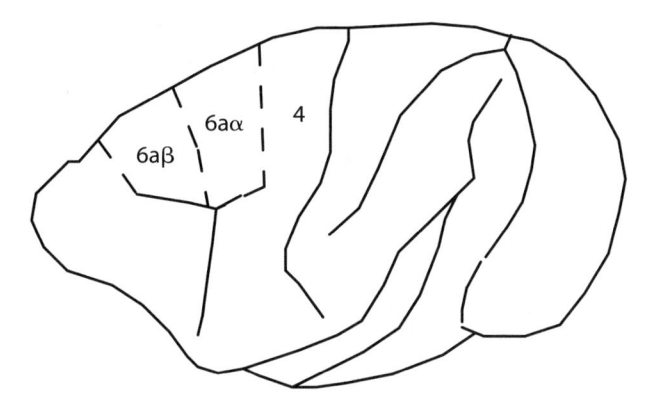

Figure 2-6 Division of the monkey motor cortex into three fields, including two dorsal premotor fields, adapted from Vogt and Vogt (1926). The scheme of Vogt and Vogt incorporated several other divisions including a set of ventral premotor areas, but their physiological work focused on the dorsal areas shown here.

separate body parts. The secondary field, 6aα, was just anterior to the primary motor field. Stimulation here at low currents evoked simple twitches similar to those evoked from primary motor cortex. However, stimulation at higher currents evoked more complex movements that combined more than one body part. The tertiary field, 6aβ, was just anterior to the secondary field. Stimulation here at low currents evoked no movement at all whereas stimulation at higher currents evoked complex movements similar to those evoked from 6aα. To the Vogts, these borders between areas were absolute and so precise that they were "hairline" divisions (Vogt and Vogt, 1926). In this respect of hard borders, the Vogts represented an extreme view. Other researchers saw at least some gradation rather than hard borders between cortical fields (e.g., Broadman, 1909; Bucy, 1935; Campbell, 1905).

To further understand the possible sequence of processing among these motor fields, Vogt and Vogt (1919) performed a set of experiments combining stimulation with fiber cutting. These experiments were not described in great detail in published form and were repeated and extended by Bucy (1933), and therefore Bucy's experiments are described here. Bucy did not distinguish between 6aα and 6aβ. His experiments involved a comparison between area 6 and the primary motor cortex, area 4. Cutting the cortex between areas 6 and 4 caused little or no effect on the movements evoked from surface stimulation of area 4. The cut, however, abolished the simple movements evoked from area 6 while leaving intact the more complex movements evoked at higher currents. Complete removal of area 4 also abolished the simple movements evoked from area 6 while leaving intact the more complex movements evoked at higher currents. In exact contrast, cutting the white matter beneath area 6, thus disconnecting its deep projections, eliminated the complex movements while leaving the simpler movements evoked from area 6. These results suggested that stimulation of area 6 evoked movements by means of two different mechanisms: the simpler movements depended on lateral connections passing into area 4, fibers that could be severed with a cut across the cortex; and the more complex movements depended on deeper connections perhaps directly to the spinal cord or to subcortical nuclei. Whether to describe 6 and 4 as linked in a hierarchical series, or whether to describe them as hierarchically equal and operating in parallel though processing different aspects of movement, was therefore ambiguous. One could find evidence of both types of relationships.

DEFINING PREMOTOR CORTEX: FULTON

It is not clear who coined the term *premotor cortex* to refer to the hypothesized higher-order motor cortex. Hines (1929) used the term in passing, but he may not have been the first. Given that *prefrontal* was already the accepted term for the anterior part of the frontal lobe, perhaps it was natural to use *premotor* to refer to the anterior part of motor cortex. In any case, the term is usually associated with Fulton who gave the name its modern connotation and popularized it (Fulton, 1934, 1935). Although Fulton performed stimulation and anatomical experiments, it was largely on the basis of lesion work in monkeys that he elaborated his concept of a premotor cortex. His starting point was an observation made by Richter and Hines in 1932.

Richter and Hines (1932) found that lesions to area 6 in monkeys caused uncontrollable grasping in the opposite hand. A monkey could even be hung from a bar by his grip. The dysfunction was temporary; after a few days the monkey appeared to regain a normal use of the hand. The deficit, however, could not be produced by a simple lesion to the cortical surface of area 6, anterior to the primary motor hand area. Instead, it depended on a massive removal of the lateral and medial surfaces of area 6 and the wedge of white matter between them. Clearly the lesion disturbed some aspect of hand function thereby causing forced grasping, but the exact function that was disturbed and its exact location in the brain were not clear.

Fulton (1935) suggested that the forced grasping produced by area 6 lesions was part of a "premotor syndrome," a loss of the highest levels of motor coordination. In his own studies, Fulton reported that lesions to primary motor cortex alone caused a temporary paralysis "most marked and enduring in the distal joints." Lesions to premotor cortex alone caused a temporary "disorganization of the more integrated voluntary movements" and some effects on visceral function. Lesions to both primary motor and premotor cortex, rather than causing a temporary deficit, caused a "permanent paralysis of voluntary movement." The monkey became permanently locked in. In Fulton's interpretation of these lesion results, the primary motor cortex controlled simple movement components whereas the premotor cortex controlled the coordination of complex motor acts. Although forming a natural motor hierarchy, the two areas were able to function partially in parallel because after lesions of the primary motor area the premotor area was evidently able to take over some of the lost function. Fulton speculated that there were two separate projection systems from cortex to the spinal cord, one from primary motor cortex commonly called the pyramidal tract, and one from premotor cortex that Fulton termed the "extrapyramidal tract."

FOERSTER

The hypothesis of a premotor cortex was even more boldly stated in the writing of Foerster (1936). During the 1920s and 1930s Foerster performed surgical operations on the brains of humans to remove epileptic foci. The patients were typically awake and under local anesthetic. Foerster was therefore able to electrically stimulate the surface of the cortex and observe the effect on behavior. Foerster (1936) stated without qualification, "The specific function of area 4 is the isolated innervation of single muscle groups" (p. 152). In contrast, when area 6 is electrically stimulated, "a complex mass movement of all parts of the contralateral half of the body is obtained" (p. 148). The border between these two areas in Foerster's map (shown in Figure 2-7) is absolute.

In Foerster's map only a type of reified myth is presented. Foerster was describing what "everybody knows." By claiming that complex and combined

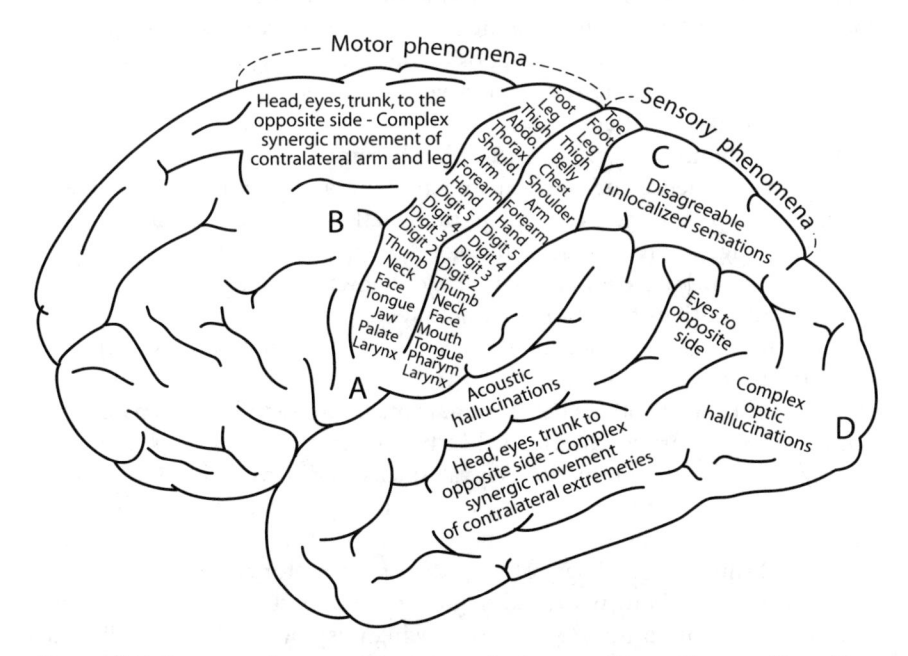

Figure 2-7 Somatotopic map of the human brain according to Foerster (from Vogt and Vogt, 1926).

movements were in the province of area 6, he was evidently forced to depict the primary motor cortex in contrast as representing only simple and separated movements. As a consequence, Foerster's primary motor cortex was an extraordinarily detailed one in which each separate fragment of the body, even each finger, had its own cortical locus. Foerster represented one of the most extreme forms of the view that the primary motor cortex is a muscle map of the body.

DOUBTING PREMOTOR CORTEX: WALSHE AND PENFIELD

Fulton's premotor cortex was not universally accepted. Walshe (1935) was highly critical of Fulton's lesion work, arguing that the "premotor syndrome" was mainly a set of vague, general symptoms that occurred with brain lesions almost everywhere, their very vagueness somehow convincing Fulton that they were "high order"; and that Fulton had merely showed the existence of a gradient in which posterior lesions in motor cortex had a larger effect, especially on the fingers, than anterior lesions.

The validity of the premotor cortex was further questioned by Wilder Penfield and colleagues (Penfield and Boldrey, 1937; Penfield and Welch, 1951). Penfield performed brain surgery on human patients under local anesthetic. To map the brain before removing diseased tissue, he electrically stimulated discrete points on the cortical surface and noted the evoked movements and also the patients' reports of induced sensation. His technique therefore was the same as Foerster's (with whom he sometimes worked), though his conclusions were in some respects the opposite. In 1937, Penfield and Boldrey published the most complete description yet of the motor map in the human brain, based on 163 patients.

Penfield's homunculus (Figure 2-8) has become stamped into the memory of every neuroscience and medical student. His name is indelibly attached to the motor map. This view of Penfield as the father of the motor map is obviously an exaggeration because he did not discover the motor map, nor was he the first to demonstrate it in humans. He did however make at least six fundamental contributions to motor cortex physiology.

First, the size of the study and the meticulous manner in which it was reported made the paper an instant classic.

Second, like many others before him, Penfield showed that movements and sensations were partially overlapped in the cortex but that movements were mainly evoked anterior to the central sulcus and sensations were mainly evoked posterior to the central sulcus. In this way he helped to establish the distinction between motor cortex and somatosensory cortex.

Third, Penfield argued against the view of the motor cortex as a roster of perfectly separated body parts. Although in his map each body part is assigned a separate location in the cortex, this drawing was not meant to be a literally accurate depiction of the motor cortex. It was meant to summarize a general trend. Penfield and Boldrey (1937) provided figure after figure of raw data

2. Early Experiments on Motor Cortex

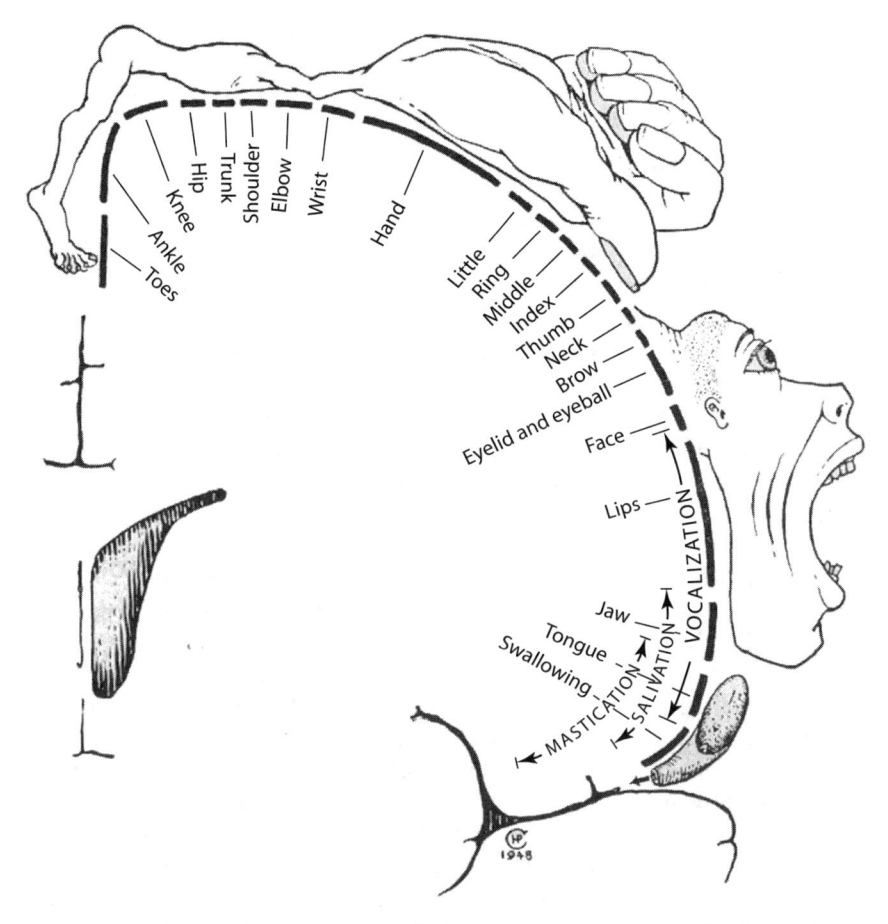

Figure 2-8 The motor homunculus of the human brain from Penfield and Rasmussen (1950). A coronal slice through the motor cortex is shown.

showing abundantly that the representations of adjacent body parts overlapped and that the map of the body was of a statistical and blurred nature. This aspect of Penfield's work is often not appreciated. He himself dismissed the novelty of the finding, pointing out that the overlap among body part representations was already obvious in the earliest studies of motor cortex (Penfield and Boldrey, 1937).

Fourth, Penfield rejected the division of motor cortex into a primary and a premotor cortex. In his view it was not the case that simple, individual movements were obtained in a distinct posterior area, or that complex, multisegmental movements were obtained in a distinct anterior area. If any distinction existed, it was a tendency for the representation of the fingers to be more concentrated in the posterior area and a representation of the trunk and shoulder to be more concentrated in the anterior area, although even this distinction was blurred. Penfield may have gone too far in this direction of denying any obvious distinction between posterior and anterior motor cortex. His position appears to have been a reaction to the summary maps of Vogt and Vogt (1919) and Foerster (1936), depicting a precise border between physiological zones, with simple movements to one side of the border and complex movements immediately to the other side. These maps appeared to be the result of imagination layered on top of a selective summary of the data. Penfield was correct in pointing out the lack of any clear border between two distinct and uniform areas. However, Penfield's interpretation that no distinctions existed, and that only one relatively homogenous motor field could be found in the lateral motor cortex, may have been too extreme and certainly caused several decades of controversy over the existence or nonexistence of the premotor cortex. To Penfield, area 4 and area 6aα belonged together into one motor cortex; and the more anterior field, area 6aβ, did not have motor functions at all. He suggested that the complex, multisegmental movements that Vogt and Vogt (1919) and Foerster (1936) had obtained from 6aβ were at least partly the result of too much current inducing an epileptic seizure that then spread indiscriminately into the motor cortex. The role of 6aβ (or PMDr as the roughly corresponding area is now called) is still debated, though in contrast to Penfield most researchers now believe that it plays at least some role in motor control.

Fifth, Penfield and Welch (1951) proposed that although there was no premotor cortex directly anterior to primary motor cortex, there was nonetheless a second motor area located on the medial part of the hemisphere. Their "supplementary motor area" (SMA) is shown in Figure 2-9. Their evidence for two motor areas was based partly on what appeared to be two somatotopic progressions of the body found in the monkey brain. In the lateral motor map, progressing up the precentral gyrus, the representation moved in a blurred fashion from the head, down the body, to the feet. In the medial motor map, beginning at the posterior part of the precentral gyrus and progressing forward, the representation moved in a blurred fashion from the feet, up the body to the head. This medial map was different from the lateral map in that it contained more overlap among adjacent body-part representations. The extent of overlap was so great that although Penfield and Welch (1951) reported the map in the monkey, they were unable to resolve the somatotopic progression in the human SMA. The medial map could also be distinguished from the lateral map by the types of movement evoked on stimulation. Stimulation of the medial map tended to evoke multisegmental movements that sometimes combined both sides of the body. Stimulation of the lateral map seemed to evoke simpler movements of one side of the body. In Penfield's interpretation, therefore, at least some of the complex actions originally localized to area 6aβ and attributed to a premotor function were more correctly identified with the SMA.

A final important contribution of Penfield's was the introduction of the term *homunculus* to describe the motor map in the human, along with a drawing of

2. Early Experiments on Motor Cortex

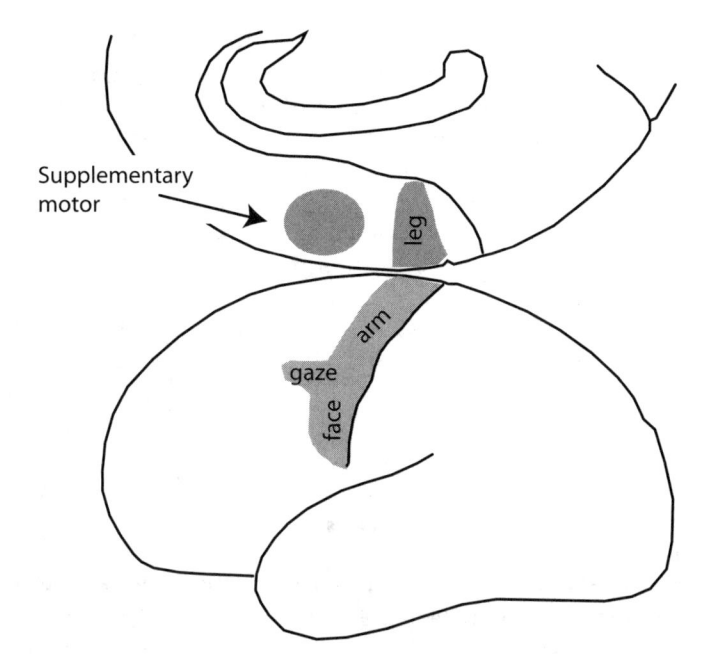

Figure 2-9 The motor and supplementary motor cortex of the human brain adapted from Penfield and Welch (1951).

the "grotesque little man" whose body proportions were distorted according to the amount of cortical representation (Penfield and Boldrey, 1937). This brilliant trick of presentation is probably the main reason for the celebrity status of Penfield's map. After the first seventy years of motor cortex research in which scientists wrote dozens of numbers and words on the brain, Penfield drew a picture; and that, proverbially, is worth a thousand words. In this case, unfortunately, the summary picture is misleading because it inevitably conveys the impression of a clean somatotopic map of the body arranged linearly on the cortex, something not present in Penfield's actual data or his written descriptions and certainly not present in the actual motor cortex.

WOOLSEY AND THE SIMCULUS

After Penfield and colleagues published the "homunculus," Woolsey et al. (1952) published a matching diagram of the monkey motor cortex that they termed a "simculus" (Figure 2-10). Following Penfield's interpretation, Woolsey depicted two motor maps, a lateral one termed "M1" and a medial one termed "M2" (Penfield's SMA). Also like Penfield, Woolsey denied the existence of Fulton's premotor cortex. The lateral motor cortex, in Woolsey's version, contained only one map of the body, the hands and feet represented in Fulton's primary motor cortex and the back and neck represented in the posterior part of Fulton's premotor cortex. The anterior part of premotor cortex, according

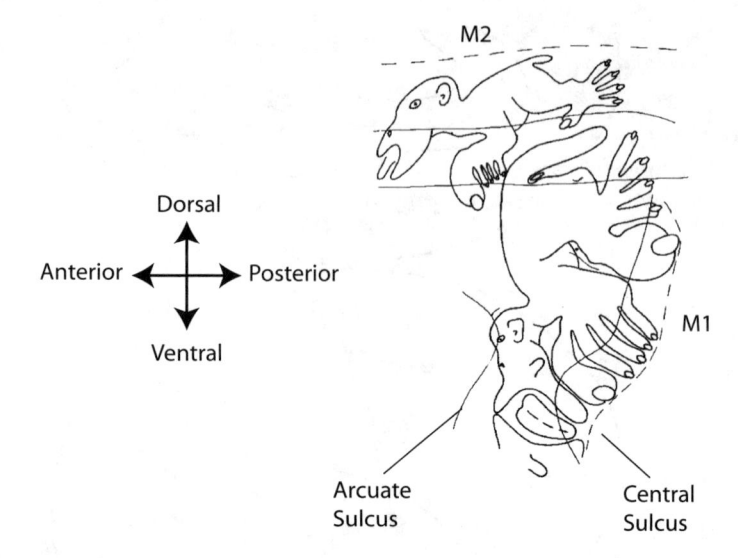

Figure 2-10 The "simculus" or map of the body in the monkey motor cortex adapted from Woolsey et al. (1952).

to Woolsey, had no motor functions. Moreover Woolsey like Penfield reported a map that was extensively overlapping. No muscle or body segment had its own separate representation.

THE INTERPRETIVE NATURE OF THE MOTOR MAPS

For eighty-two years, from Fritch and Hitzig (1870) to Woolsey et al. (1952), scientists applied electrical stimulation to the surface of the motor cortex. Remarkably, using more or less the same technique over nearly a century, different researchers saw very different patterns. The organization of motor cortex is apparently something of a Rorschach inkblot and invites interpretation.

Originally only one map of the body was recognized, a blurred map with considerable overlap among adjacent representations. During the early twentieth century, a new view emerged of two motor areas linked in a hierarchy, a premotor cortex with no discrete topography, and a primary motor cortex with a discrete map of the body (Campbell, 1905; Foerster, 1936; Fulton, 1934, 1935; Vogt and Vogt, 1919, 1926).

Penfield and colleagues (Penfield and Boldrey, 1937; Penfield and Welch, 1951) and Woolsey et al. (1952) proposed a different organization to motor cortex. In their view, the premotor cortex did not exist. Instead, motor cortex contained two somatotopically organized areas, a lateral one (M1) and a medial one (SMA or M2).

Penfield, Woolsey, and colleagues believed that they had arrived at a definitive description. Yet the integrity of SMA as a distinct area should not be exaggerated. Its borders are blurred with the surrounding motor cortex. Horsley and Schaffer (1888) had already accurately described the same pattern, including the somatotopy on the medial wall of the hemisphere but interpreted it differently, drawing the borders such that a single continuous map of the body could account for the lateral and the medial motor cortex.

Perhaps one reason for this diversity of interpretations is that the topographic organization of motor cortex is fundamentally multiply determined. As detailed in Chapter 10, one hypothesis is that there is no single scheme or division into sectors or body maps that can explain all the complexities of the pattern. Instead a deeper principle may be at work, in which the many dimensions of the movement repertoire are rendered onto the two-dimensional sheet of the cortex while optimizing local continuity (Aflalo and Graziano, 2006b). In this view, the overlaps, swirls, fractures, partial separation into functionally distinct areas, and other confusing features of the topographic arrangement can be explained by the interactions and compromises among different mapping requisites.

BEYOND SURFACE MAPPING

The maps of Penfield and colleagues and of Woolsey and colleagues represent the end of a line of research in which motor cortex was mapped through surface stimulation. Little more was to be gained by continued surface mapping. Instead, subsequent experiments used more refined techniques to pursue three broad questions:

- Somatotopy: What is the extent of somatotopic overlap in the lateral motor cortex, and does it change with experience?
- Hierarchy: To what extent can motor cortex be divided into many separate areas that process movement at different levels of abstraction?
- Single-neuron properties: Is it possible to understand the specific mechanism of movement control by monitoring the activity of single neurons in the motor cortex of an experimental animal as it performs trained movements?

These three broad questions account for much of the research on motor cortex from the 1950s to the present. They are reviewed in Chapters 3–5.

Chapter 3

An Integrative Map of the Body

INTRODUCTION

A traditional view of the primary motor cortex is that it decomposes movement into individual joint or muscle actions. The evidence, however, is overwhelmingly against this ubiquitous view. There is apparently no place in cortex where muscles, joints, or body parts are separated into strictly independent control elements. The maps are more integrative than decompositional.

Many of the experiments that addressed this question of somatotopic overlap focused on the control of the hand muscles. This chapter first discusses the question of whether the musculature of the hand is controlled in a decomposed or in an integrated manner in primary motor cortex. The chapter then discusses integration among other body-part representations, and how this integration may develop through experience with complex actions.

ARE THE MUSCLES OF THE HAND OVERLAPPED IN THE PRIMARY MOTOR CORTEX?

Asanuma and the Proposal of Cortical Columns for Individual Muscles

As summarized in Chapter 2, the first seventy years of research on motor cortex involved primarily surface stimulation. The resulting maps suggested an overlapping somatotopy. Different body parts appeared to be controlled by partially overlapping regions of cortex (e.g., Penfield and Boldrey, 1937; Woolsey et al., 1952; see also Landgren et al., 1962). The overlap, however, might have been caused partly by the spread of electrical current over the cortical surface, blurring a discrete underlying map. To reduce the problem of current spread, Asanuma and colleagues mapped the motor cortex in cats (Asanuma and Sakata, 1967; Asanuma and Ward, 1971) and the primary motor cortex in capuchin monkeys (Asanuma and Rosen, 1972) with the technique of intracortical microstimulation.

A microelectrode, a fine hair-like wire sharpened at one end and insulated except at the tip, was inserted into the cortex, allowing the experimenters to electrically stimulate a small sphere of tissue surrounding the electrode tip. The stimulation consisted of negative-going pulses of current because negative pulses were known to discharge neurons more effectively than positive pulses. Each pulse was a fraction of a millisecond in duration (0.2 ms). The amplitude was usually set to a low value, typically 20 microamps or lower (This current was certainly much lower than the current used, for example, by Fritsch and Hitzig [1870/1960]; they described their current level as just enough to be felt as a tingle when applied to the experimenter's tongue.). Pulses were presented in rapid succession (for example at a rate of 300 Hz), and the entire train of pulses was typically less than 30 ms in duration though sometimes stimulation as long as 1 second was used. This stimulation train applied inside the cortex rather than to its surface evoked a muscle twitch that the experimenters could easily observe. It also evoked neuronal activity in peripheral nerves that the experimenters could directly measure.

This technique of microstimulation was not without caveats. Because of the spatial focus of the current, stimulation trains longer than half a second, or currents higher than about 20 microamps, tended to kill the local brain tissue with an accumulation of negative charge at the electrode tip (Asanuma and Arnold, 1975). Asanuma and colleagues reduced this problem through a variety of tricks such as regularly discharging the electrode, but the problem can be entirely solved by applying biphasic pulses rather than negative pulses (e.g., Tehovnik, 1996). A biphasic pulse has first a negative and then a positive deflection, balancing the charge and allowing even very high currents and long stimulation trains to be used without killing brain tissue. Asanuma's use of negative pulses became standard in motor cortex physiology and is still often used, forcing experimenters to limit the upper range of their stimulation parameters.

Asanuma's results led him to suggest that the wrist and hand representation in the primary motor cortex was organized into a mosaic of patches or columns, each column about half a millimeter wide (Asanuma, 1975). In most cases, in his interpretation, a cortical column was connected to a single muscle, and therefore stimulation of that column caused a flexion or an extension of a single joint. In a minority of cases a column might be connected to more than one muscle. Asanuma's view is to date the most extreme form of a map of muscles. No other major investigator suggested a discrete representation of individual muscles.

Asanuma's interpretation of single-muscle columns in cortex was derived from his use of threshold stimulation. In this technique, the stimulating current was adjusted until the muscle activity was only just detectable to the experimenters. The threshold for most sites in the primary motor cortex was less than 20 microamps. For sites in the hand representation where Asanuma reported single-muscle columns, the threshold was usually lower than 10 microamps and could be as low as 3 microamps. Although stimulation above threshold often activated complex ensembles of muscles, at threshold Asanuma typically observed activation of only one muscle or rotation of one joint. Asanuma (1975) was careful to point out that these results may pertain specifically to the control of the finger muscles, and moreover to a specific posterior zone in the monkey primary motor cortex in which the thresholds for the finger muscles were especially low. He speculated that the individual and detailed nature of finger movements may require a machinery in which

3. An Integrative Map of the Body 41

each muscle is individually controlled by cortex. In support of this speculation, he noted that control of the wrist muscles in the monkey did not appear to be as segregated into nonoverlapping columns for individual muscles.

Asanuma's results using threshold stimulation, however, were susceptible to a second interpretation. In any normal movement, some muscles are more active than others, and one or two muscles may be most active of all. Most normal, behaviorally useful movements do not involve equal, simultaneous activation in all participating muscles. If a point in cortex represents a complex movement, then at threshold stimulation, by definition most of the components of the movement are no longer visible and only the one or two strongest components remain observable. In this tip-of-the-iceberg hypothesis, lowering the current until the movement is barely detectable will always result in a simpler movement because it allows only the tip of the movement iceberg to be expressed. Asanuma's results from threshold stimulation were consistent with the hypothesis of cortical columns for single muscles, but also with the tip-of-the-iceberg hypothesis. The data therefore did not resolve the question of whether single muscles were represented in discrete cortical regions.

Motor Cortex Neurons Correlated with Many Muscles

The question of whether muscles are discretely represented in cortex was answered unambiguously in a set of experiments by Cheney and Fetz (1985). They used microelectrodes to record the naturally occurring activity of single neurons in the primary motor cortex of monkeys. At the same time, they recorded the activity of eight muscles actuating the wrist and fingers. When a single neuron in cortex fired an action potential, it caused a minute effect on the activity of the muscles. This effect was so subtle that it could be seen only by averaging the results of thousands of neuronal spikes. This technique of spike-triggered averaging showed that a single neuron in cortex could affect muscles in the arm with a latency as short as 5 ms. Clearly, Cheney and Fetz had tapped into the most direct, descending pathway from cortex to the muscles. This pathway must have included cortical neurons projecting to the spinal cord, perhaps interneurons in the spinal cord, and motoneurons in the spinal cord projecting to the muscles. Each neuron in cortex was found to relate to a set of muscles, not just to one muscle. The spiking of one neuron in cortex might be followed by a complex pattern of varying degrees of excitation in one set of muscles and inhibition in another set of muscles, as if the function of the neuron was to participate in a complex, multimuscle action. These patterns of muscle activity varied from neuron to neuron. Some neurons were linked to coactivation of the flexors and the extensors of a joint; other neurons were linked to activation of the flexors and inhibition of the extensors, or vice versa; some neurons influenced the muscles that crossed many joints; others were limited mainly to one joint. Presumably the eight muscles measured in the experiment provided a limited window on what must have been an even more widespread pattern of muscle activity. Clearly the mapping from cortical neurons to muscles was not one-to-one but instead was many-to-many.

The experiment described above involved recording the natural activity of neurons, not artificially stimulating them. Do the properties of neurons, as determined in a recording experiment, match the properties determined by electrically stimulating the same site in cortex? Cheney and colleagues (Cheney and Fetz, 1985; Cheney et al., 1985) examined this question using the technique of "stimulus triggered averaging" in which stimulation pulses were applied to cortex. The pulses were presented with a long enough interpulse interval (66 ms) that each pulse could be considered approximately an independent event. Each pulse applied to cortex evoked a minute effect at the muscles. The results of many thousands of pulses were then averaged. The mean effect of a stimulation pulse closely matched the results from the single-neuron analysis. A similar pattern of excitation of some muscles and inhibition of other muscles was evoked, whether by a stimulation pulse applied to a site in cortex, or by the spiking of a single neuron in the same site in cortex. The stimulation presumably directly activated a set of neighboring neurons that shared similar output properties.

The results of the experiments of Cheney and colleagues provided a conclusive answer to a basic question. Muscles are not typically represented in isolated patches in the primary motor cortex. Each location in cortex, and even each individual neuron, when active, evokes a complex pattern of activity across a set of muscles. The results strongly support the "tip-of-the-iceberg" hypothesis. Asanuma's apparent columns for single muscles (Asanuma, 1975) must have been the result of lowering the stimulation until only the tip of the movement iceberg was detectable.

Further Studies of Overlapping Representations of the Fingers

The overlap in the cortical representation of fingers was further studied by Schieber and Hibbard (1993), who trained monkeys to make flexion and extension movements of individual fingers. As the monkey performed the task, the experimenters recorded the activity of single neurons in the hand representation in the primary motor cortex. They found that the majority of neurons fired in relation to the movement of more than one finger. The fingers were therefore represented in an overlapping manner in cortex. These results were consistent with the findings of Cheney and Fetz (1985) and in contradiction to Asanuma's (1975) interpretation of separate cortical columns for individual muscles that flex or extend the fingers.

Rathelot and Strick (2006) traced the anatomical connectivity in monkeys between the motor cortex and the muscles of the hand. The experiment involved injection of rabies virus into a hand muscle. The virus spread up the axon of the motor neuron into the spinal cord, crossed a synapse, and spread through the secondary neuron. The animal was sacrificed before the virus had time to spread to a tertiary neuron. For this reason, the method labeled only those cortical neurons that projected directly to the motor neurons of the injected muscle. The population of neurons in cortex projecting to a single hand muscle was widespread, covering a large area of the primary motor cortex, and the populations of neurons projecting to different hand muscles were fully overlapping. These results further confirm that the representations of the finger muscles are intermingled in the monkey motor cortex and not separated into columns.

The overlap of finger representations was also demonstrated in the human primary motor cortex using functional magnetic resonance imaging (fMRI) to measure blood flow in the hand representation while the participants performed individual finger movements (Sanes et al., 1995). Subsequent imaging studies suggested that even though the finger representations are overlapped in the human primary motor cortex, each finger may have a cortical hot spot within the larger region of overlap, the hot spots arranged in a somatotopic sequence with the thumb represented laterally and the pinky represented medially (Beisteiner et al., 2001; Dechent and Frahm, 2003; Kleinschmidt et al., 1997).

It should be noted that there is a distinction between overlap among muscle representations and among finger representations. The fingers are mechanically linked. To flex one finger without the others is not a matter of contracting one muscle; it requires the coordination of muscles that move or stabilize all the fingers. In this sense, therefore, the hypothesis of a muscle-by-muscle map and the hypothesis of a finger-by-finger map are mutually inconsistent. A map in which each finger has a segregated cortical representation would be a map of complex movement that integrates the action of many muscles. The representation in primary motor cortex of humans appears to be partially separated in this manner; there is some partial or relative separation into individual finger representations. However, these finger representations are extensively overlapped.

SOMATOTOPIC OVERLAP BEYOND THE FINGERS

The studies described above focused on the mapping of hand muscles in the primary motor cortex. To what extent is the hand represented separately from the shoulder and elbow, and to what extent is the arm represented separately from other body parts?

Kwan et al. and the Core-Surround Organization

Kwan et al. (1978) mapped the arm representation in the primary motor cortex of monkeys using microstimulation and were the first to report the nested organization shown in Figure 3-1. In a posterior, core zone of cortex, stimulation caused movements of the fingers and wrist. This core zone was clearly the same as the finger region that Asanuma had studied (Asanuma, 1975; Asanuma and Rosen, 1972). In a zone of cortex partially surrounding the core, stimulation caused movements of the elbow and shoulder. This curious orga-

ELBOW MOVEMENTS   FINGER MOVEMENTS

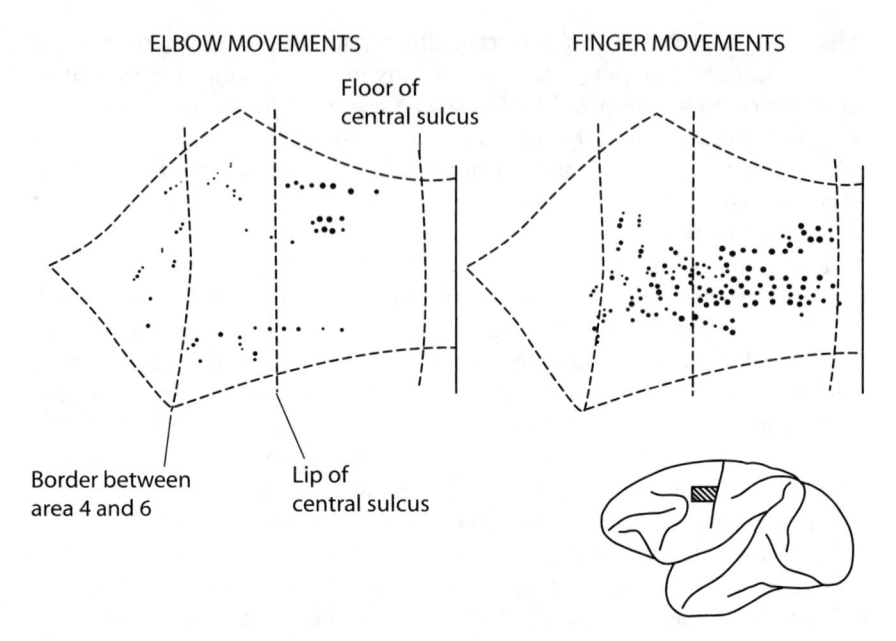

Floor of central sulcus

Border between area 4 and 6

Lip of central sulcus

Figure 3-1 The nested organization of the monkey motor cortex adapted from Kwan et al. (1978). The dots indicate cortical sites from which stimulation elicited movements of the elbow (left panel) or the fingers (right panel). The shaded rectangle on the macaque brain drawing (lower right) indicates the approximate location of the studied cortex

nization was not quite like the Penfield homunculus, with its vertical stacking of body parts in which the arm is dorsal to the hand (see Figure 2-8). It was similar though not quite the same as the map of Woolsey et al. (1952), with the hand in a posterior region and the arm in an anterior region (see Figure 2-10). This wrapping of the arm representation around the finger representation departs from the common textbook description but has now been consistently observed. It has been reported by others in the monkey motor cortex (Park et al., 2001; Sessle and Wiesendanger, 1982) and in a recent experiment in the human motor cortex (Meier et al., 2007).

In interpreting the Kwan et al. result, it is essential to remember the "tip-of-the-iceberg" caveat of stimulation at threshold. The data show that the core region emphasizes the fingers more than the arm; that the belt region emphasizes the arm more than the fingers; and that it is possible to lower the current and reduce the strength of the movement until an apparent separation is achieved. An absolute separation between the hand representation and the arm representation cannot be inferred or refuted from that data.

Other Maps of the Hand and Arm Representations

Gould et al. (1986) published a microstimulation map of the lateral motor cortex of owl monkeys, again focusing on the hand and arm. Their map showed

3. An Integrative Map of the Body

even more clearly a lack of any simple stacking of body parts. They reported nearly a chaos of local patches of cortex representing different body parts. The fingers were represented in multiple patches, not merely in one core zone or hot spot. The authors depicted the motor cortex as a mosaic of interlocking zones, each zone representing a single body part. Yet again, one must be wary of the iceberg caveat. Each cortical patch may have represented a complex, multimuscle movement, and the threshold stimulation may have evoked only the tip of that movement iceberg. For example, a "forearm" zone may have represented a complex movement of which the forearm happened to be the strongest component and therefore by definition the only one still detectable at threshold.

Donoghue et al. (1992) published yet another map of the monkey primary motor cortex, using microstimulation while measuring the activity of a sample of fourteen muscles from the arm and hand. They reported a greatly overlapping organization. Stimulation of most cortical sites evoked activity in more than one muscle. The representation of the hand muscles and the upper arm muscles were extensively overlapped.

Park et al. (2001) revisited the question of somatotopy in the arm and hand representation of the monkey primary motor cortex using the technique of stimulus triggered averages. They applied stimulation pulses to the cortex and measured the activity in a range of arm and hand muscles. Their maps beautifully confirm the nested organization first described by Kwan et al. (1978). Park et al. (2001) found three zones shown in Figure 3-2: a core region on the posterior edge of the primary motor cortex, stimulation of which evoked activity in hand muscles; a surrounding region, stimulation of which evoked activity in the hand and arm muscles; and a larger encompassing belt, stimulation of which evoked activity in the arm muscles only. Once again this study must be interpreted in light of the iceberg caveat. Strictly speaking, the results show that the distal muscles are more strongly represented in the core area, such that near threshold only distal muscles are activated; proximal muscles are more strongly represented in the outer belt area, such that near threshold only proximal muscles are activated; and distal and proximal muscles are represented more equally in the intermediate belt area, such that threshold stimulation cannot easily separate the two outputs.

More recent mapping data from the same lab (Park et al., 2004) obtained a similar result with twenty-four recorded muscles. This study showed that almost all cortical sites were related to muscles from more than one joint. About half the sites were related to hand and upper arm muscles, and many sites were related to every joint in the arm.

In our own stimulation studies on monkeys (Graziano et al., 2005; Graziano, Taylor, et al., 2002), when we stimulated with trains of electrical pulses extended to a behaviorally relevant time scale, we also obtained a core and surround organization. In the core area, stimulation evoked movements resembling manipulation or grasp of objects. As a result, these movements emphasized the wrist and fingers with relatively less involvement of the arm.

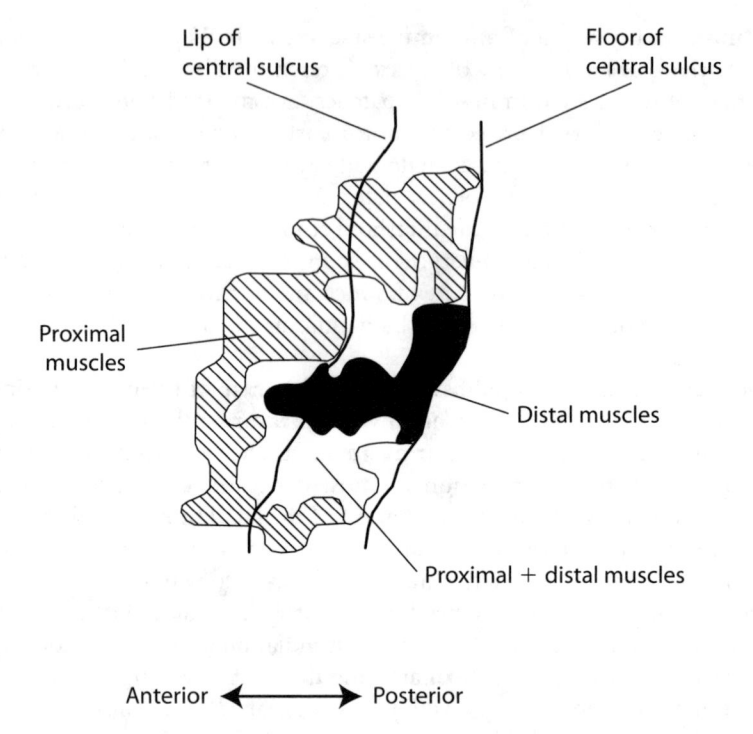

Figure 3-2 The nested organization of the monkey motor cortex adapted from Park et al. (2001). Movements of the distal muscles (fingers) were evoked by stimulating in a core area, movements of the proximal muscles (arm) were evoked by stimulating in a surrounding area, and combined movements of distal and proximal muscles were evoked from an intermediate belt.

Stimulation in the motor cortex surrounding the core area evoked a variety of arm movements such as apparent reaches into different areas of space. These movements emphasized the arm and had relatively less involvement of the fingers. The core and surround organization was therefore a relative one. The separation between arm and hand was not absolute. Moreover, the combined movements of the arm and the hand evoked by stimulation seemed to be coordinated, resembling fragments of the normal behavioral repertoire. It appeared, therefore, that the control of the arm and of the hand was not segregated in primary motor cortex but instead was elaborately integrated for the purpose of controlling meaningful behavior.

Lack of Overlap Among Major Body Segments in Primary Motor Cortex

Almost all studies to address the question of somatotopic overlap in the primary motor cortex have focused on the hand and arm representation. In this area of cortex, as reviewed above, the muscles of the hand and arm have extensively overlapped representations. Relatively few studies have examined the overlap of major body segments such as the arm and face, or the arm and leg. The few studies that exist, however, tend to suggest that there is minimal overlap across major body segments. For example, Huntley and Jones (1991) used anatomical tract-tracing techniques to study the lateral connectivity among neurons in the primary motor cortex of monkeys. They found that within the arm and hand area, neurons were densely laterally connected, suggesting an extensive somatotopic integration. Likewise, within the face area, neurons were densely laterally connected. However, few lateral projections connected the arm representation to the face representation, suggesting a relatively clean separation between the major body segments.

A more recent experiment (Park et al., 2004) used electrical stimulation to map the primary motor cortex of monkeys and confirmed that the arm and hand representation contains extensive overlap among muscle representations and yet is relatively discretely separated from the face and the leg representation.

The studies reviewed above suggest that the primary motor cortex does not decompose movement into separate muscle contractions or joint rotations. It is more integrative than decompositional. Yet there is a limit to the integration. It appears to have a relatively discrete separation among the major body segments.

INCREASE IN OVERLAP WITH EXPERIENCE

The amount of somatotopic overlap within the primary motor cortex is not fixed but instead can change with behavioral experience. For example, Nudo et al. (1996) found that monkeys that practice the combined use of two arm joints develop greater overlap in the cortical representation of those joints.

Martin and colleagues (Chakrabarty and Martin, 2000; Martin et al., 2005) further explored the role of experience in the development of an overlapping somatotopy. They used microstimulation to map the motor cortex in cats. They found that at birth, the representation in motor cortex was mainly nonoverlapping. Separate joints of the forelimb were represented in discrete patches in cortex. During development, as the kitten learned to perform complex behaviors that required coordination among joints, the representations in cortex developed the highly overlapped property characteristic of the adult. Individual joints were no longer typically represented in separate patches. If the kitten was prevented from practicing complex, integrated movements, the motor map did not develop the normal overlap of representations.

These results suggest that during experience the motor cortex is trained on, and comes to reflect, the movement repertoire of the animal. If an animal has a need to control individual muscles (if such an unlikely condition ever exists), the animal might well develop a motor cortex map that topographically separates the muscles. In the more common case that an animal has a need to control many muscles and joints in a coordinated fashion, such as for reaching toward an object or manipulating an object, its motor cortex develops a topography in which the relevant muscles are represented in an integrated fashion.

DOES SOMATOTOPIC OVERLAP REFLECT HIERARCHY OR BEHAVIORAL REPERTOIRE?

The results reviewed above suggest that a discrete map of muscles exists nowhere in cortex, that all motor cortex areas contain somatotopic overlap, and that different areas have different amounts of somatotopic overlap.

In a traditional interpretation, the SMA stands at the highest level of a processing hierarchy, integrating movements across the entire body bilaterally. The lateral premotor cortex stands at an intermediate level, coordinating among body parts but not to the same extent as the SMA. The primary motor cortex stands at the lowest level, controlling individual movement components. This simple hieraerchical model, however intuitively appealing, has serious difficulties. Whether it is flat out wrong, or simply requires some nuancing, is not yet clear. Three aspects of the data do not fit easily with the hierarchical model.

First, as reviewed above, the primary motor strip does not control individual muscles or joints. It contains a highly integrative map more consistent with the control of meaningful behavior, albeit behavior focused on to one or another major body segment.

Second, according to the hierarchical view, the primary motor strip and the premotor strip should control the same movement repertoire at different hierarchical levels. The primary motor strip, however, has a relative emphasis on the distal body parts, including the hands, fingers, feet, and toes, whereas the premotor strip contains a relatively greater emphasis on the trunk, neck, and shoulder. The hierarchical view cannot easily explain why the two regions emphasize different parts of the body.

Third, the primary motor cortex, the premotor cortex, and SMA are not connected in a strict series, but instead all have direct outputs to the spinal cord. Projections to the spinal cord from the premotor areas were first suggested by the physiology studies of Vogt and Vogt (1919) and Bucy (1933) who showed that stimulation of premotor cortex could evoke movements even after the primary motor cortex was removed or undercut. Over the subsequent decades a large number of studies using anatomical tracing techniques verified this direct projection from the premotor areas to the spinal cord (e.g., Dum and Strick, 1991; Galea and Darian-Smith, 1994; Hoff and Hoff, 1934; Kennard, 1935; Kuypers and Brinkman, 1970; Macpherson, Macpherson, Wiesendanger et al., 1982; Murray and Coulter, 1981; Nudo and Masterton, 1990; Toyoshima and Sakai, 1982).

An alternative to the hierarchical model was suggested by our electrical stimulation experiments. Stimulation of different zones within the motor cortex evoked different kinds of movements from the animal's normal repertoire. For example, stimulation of SMA evoked movements resembling leaping or climbing. This type of behavior obviously requires bilateral integration of the trunk and limbs, perhaps explaining the extensive bilateral somatotopic integration in SMA. Stimulation of the primary hand area evoked movements resembling the manipulation of objects already in grasp. This behavior requires

3. An Integrative Map of the Body

extensive integration of muscles of the hand, some involvement of the arm and shoulder, and little involvement of the rest of the body, neatly matching the pattern of somatotopic overlap found in the primary motor hand area. The stimulation experiments therefore suggested that different cortical zones were distinct from each other mainly in the category of behavior that they emphasized.

In summary, there are now two possible theoretical explanations for the heterogeneity in the motor cortex. A traditional explanation is that the different cortical motor areas represent different levels in a processing hierarchy. A second explanation, suggested by our stimulation experiments, is that the different cortical areas represent statistical clusters in the movement repertoire. We have leaned toward this second explanation, even so far as to suggest that the "primary" motor cortex is no more primary than the premotor cortex or the SMA. Admittedly, however, in proposing a new idea one may tend to kick the old idea too hard. It is useful to keep in mind that the hierarchical explanation and the behavioral-cluster explanation are not necessarily incompatible and might both be valid. These questions are addressed in greater detail in the next chapter.

Chapter 4

Hierarchy in the Cortical Motor System

INTRODUCTION

Figure 4-1 shows a proposed hierarchical scheme for the cortical motor system. This scheme contains three general classes of cortical motor area. First is a mosaic of output areas. These cortical "action zones" are organized around the broad categories of action that make up the animal's behavioral repertoire. They correspond to the traditional primary motor cortex, caudal premotor cortex, SMA, and possibly the cingulate motor areas, all of which, in the present scheme, are at approximately the same hierarchical level, and all of which project directly to the spinal cord. The exact connectional patterns and functional properties vary among these output areas according to the requirements of the actions emphasized within them. For example, a cortical region that emphasizes defense of the body surface may require specific input carrying visual information about nearby objects. A cortical region emphasizing manipulation of objects may require direct output to spinal motor neurons for the dexterous control of the hand and fingers. A second class of motor area in this proposed scheme is a set of parietal areas that provides a liaison between sensory processing and motor output (for a similar point see also Matelli and Luppino, 2001; Wise et al., 1997). A third class of motor areas is a set of rostral premotor areas that provides a liaison between the prefrontal areas and the output areas (for a similar point see Lu et al., 1994; Takada et al., 2004).

The difference between the scheme proposed here and the traditional view lies mainly in the cortical output areas. In the traditional view, the main cortical output is a single map of muscles in the primary motor cortex. That map represents individually meaningless movements that higher-order areas can combine into meaningful actions. In the modified scheme described here, many output zones exist, each one emphasizing a different meaningful action category.

These proposed cortical action zones are not strictly separate areas. For this reason they are drawn schematically as overlapping ovals in Figure 4-1. They are more like clusters. They are hills that emerge with different movement emphases.

The output zones are also not strictly on the same hierarchical level. For this reason they are depicted at different heights. Broadly speaking, they are part of the cortical output. Yet they emphasize movements with very different control requirements. It is likely that among the output zones are differences in complexity, in the level of abstraction of the information that is processed,

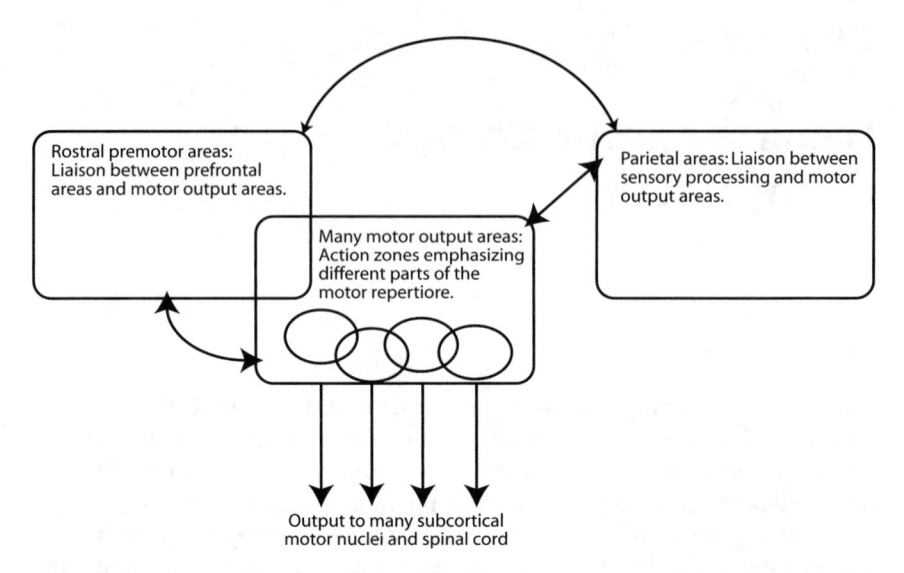

Figure 4-1 Proposed hierarchy of the cortical motor system.

and in the manner in which information flows laterally from one zone to another. For these reasons, it is probably not correct to think in terms of rigid hierarchies with absolute stages.

The hypothesis of hierarchy in the cortical motor system has a long history full of conflicting opinions. Whether any premotor cortex existed was initially controversial. Researchers then began to describe not one but at least six premotor areas. The exact properties of these areas are still debated. That they are different from each other is clearer than what they do or how they relate to each other. The following sections review the emergence of ideas on these many premotor areas. The chapter ends by returning to the hierarchical model proposed in Figure 4-1 and discussing some of its nuances and caveats in greater detail.

PENFIELD AND WOOLSEY DENY PREMOTOR CORTEX

As described in Chapter 2, in 1905 Campbell proposed that the motor cortex could be divided into a higher order, anterior part and a primary, posterior part (see Figure 2-5). Fulton (1934, 1935) championed this view of a primary motor and premotor cortex. Penfield and Welch (1951) and Woolsey et al. (1952) proposed an alternative organization. In their formulation, the correct division lay between the lateral motor cortex that contained one map of the body (M1), and the medial motor cortex that contained a second map of the body (M2 or SMA) (see Figure 2-10.)

The views of Penfield and of Woolsey were profoundly influential. No one has ever seriously questioned the existence of SMA as a separate motor area.

Furthermore, the hypothesis of a premotor cortex distinct from the primary motor cortex was, on their suggestion, abandoned and not seriously reconsidered for thirty years. Not until the 1980s did research begin to accumulate again for a distinction between the primary motor and premotor cortex.

EARLY IMAGING OF THE HUMAN MOTOR CORTEX: ROLAND AND COLLEAGUES

Roland and colleagues were among the first to image the activity of the human brain during the performance of a task (Roland and Larsen, 1976; Roland, Larsen, et al., 1980; Roland, Skinhoj, et al., 1980). They measured regional cerebral blood flow by injecting a radioactive tracer into the carotid artery of participants and then measuring the radiation via an array of cameras. They asked their participants to perform a variety of tasks in an attempt to distinguish among the primary motor cortex, premotor cortex, and SMA.

When participants were asked to manually palpate an object, cerebral blood flow increased in the primary motor and somatosensory cortex (Roland and Larsen, 1976). This activity could be separated into its sensory and motor components. Passive somatosensory stimulation of the hand caused an increase of blood flow mainly in the primary somatosensory cortex, whereas active palpation of an object, when sensation in the hand was blocked by local anesthesia, caused an increase of blood flow mainly in the primary motor cortex.

When participants were asked to tap the ball of the thumb against the ball of the other four fingers in a fixed sequence, cerebral blood flow was particularly pronounced in SMA (Roland, Larsen, et al., 1980). SMA was activated by either hand, confirming its relationship to bilateral movement.

When participants were asked to point to squares on a grid, moving the index finger from square to square on the basis of ongoing verbal instructions, cerebral blood flow was particularly pronounced in the premotor cortex (Roland, Skinhoj, et al., 1980).

On the basis of these results, Roland and colleagues suggested that the primary motor cortex participated in the execution of motor tasks; SMA participated in the coordination of motor sequences, or the internal triggering of movements; and premotor cortex participated in the establishment of new motor programs or the modification of previous ones.

Figure 4-2 shows Roland's parcellation of the human brain into functional areas. Premotor cortex according to Roland was in a dorsal location, just anterior to the primary motor arm representation. It roughly corresponded to the dorsal premotor areas 6aα and 6aβ described by Vogt and Vogt (1919, 1926), the dorsal premotor area of Fulton (1934, 1935), and the premotor area of Foerster (1936). In one respect, however, Roland's parcellation was different from most previous ones. Roland correctly placed an eye movement area, the FEF, in the middle of human area 6, directly anterior to the lateral motor strip. Modern maps of the FEF show a larger zone in a similar location (e.g., Kastner et al., 2007). The FEF occupies much of the precentral gyrus. This placement

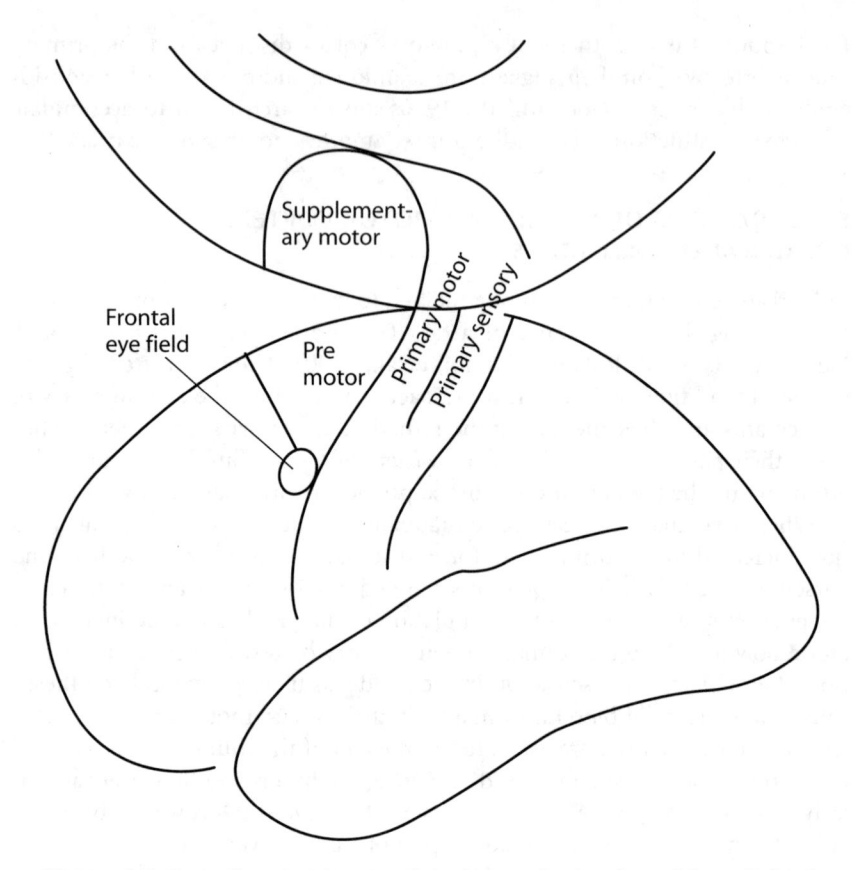

Figure 4-2 Some functional areas of the human brain adapted from Roland, Larsen, et al. (1980).

of the FEF requires that the human premotor cortex be squeezed either above the FEF (as Roland drew it) or above and below it.

FRACTIONATION OF PREMOTOR CORTEX INTO SEPARATE FIELDS: RIZZOLATTI AND COLLEAGUES

Arguably nobody has done more to establish the organization of premotor cortex than Rizzolatti and colleagues. Most of their work has focused on the monkey motor cortex. They presented several lines of evidence to argue that the lateral premotor cortex exists as a separate motor area anterior to the primary motor cortex, and that it controls movement at a higher level of abstraction. They also proposed that the lateral premotor cortex is not a unitary area but is divisible into at least four subareas that participate in different though not fully understood aspects of movement control.

Matelli et al. (1985) examined the pattern of staining for cytochrome oxidase in the motor cortex of monkeys. Cytochrome oxidase is a mitochondrial

4. Hierarchy in the Cortical Motor System

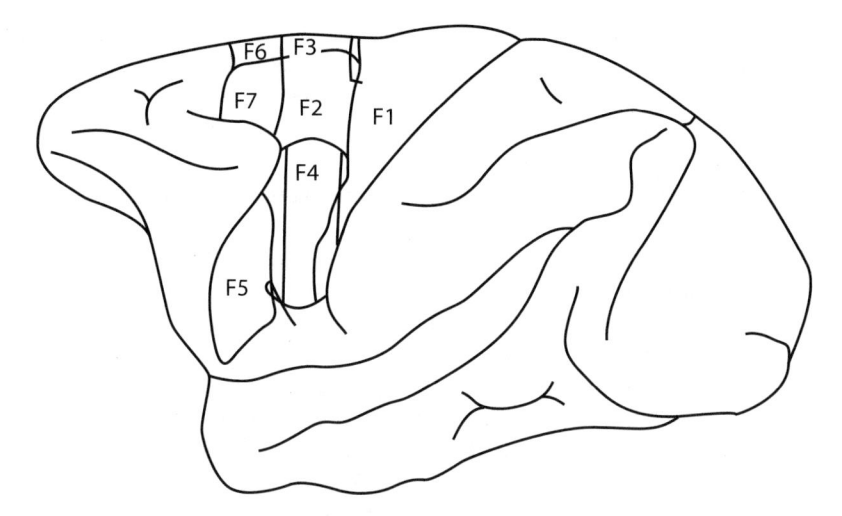

Figure 4-3 Some divisions of the monkey motor cortex based on the histological work of Matelli et al. (1985, 1991).

enzyme more prevalent in cells that are more metabolically active. The stain reveals subtly different patterns in different regions of cortex. Matelli et al. reported five distinct motor zones or "fields" in the monkey, shown in Figure 4-3. Field 1 (F1) roughly corresponded to the primary motor cortex; F2 to dorsal premotor cortex; F3 to the SMA; and F4 and F5 to the ventral premotor cortex. Later, the same group further subdivided the dorsal premotor cortex into a posterior subregion that they labeled "F2" and an anterior subregion that they labeled "F7" (Matelli et al., 1991). The lateral premotor cortex, therefore, was divided into quadrants: dorsal anterior (F7), dorsal posterior (F2), ventral anterior (F5), and ventral posterior (F4). This parcellation of the lateral premotor cortex is now commonly accepted, although the terminology varies among investigators.

THE VENTRAL PREMOTOR CORTEX

Multiple Hand Areas

Rizzolatti and colleagues focused their physiological work on the ventral portions of premotor cortex, their F4 and F5, and on the adjacent primary motor cortex, their F1. They recorded the activity of single neurons while the monkey made spontaneous movements or was tested with passive somatosensory stimuli. They reported two distinct hand representations, one in primary motor cortex and one in F5 (Gentilucci et al., 1988; Rizzolatti et al., 1988). In the primary motor hand representation, the neuronal responses were related primarily to the fingers and wrist. Just anterior to this primary motor hand representation, neurons in F4 were related more to movements of the upper arm and trunk rather than to the hand. Just anterior to F4, the neurons in F5 were once again related to the fingers and wrist, responding especially during grasping of objects with the hand and interactions between the hand and the mouth. This finding of a progression from distal musculature in the primary motor cortex, to proximal musculature in F4, and returning to distal musculature in F5 provided strong evidence that the lateral motor cortex did not contain a single, simply organized map of the body.

Strick and colleagues have gathered evidence that the lateral motor cortex contains at least three hand areas: one in traditional primary motor cortex, one in the ventral premotor cortex, and one in the dorsal premotor cortex, all three of which project to the spinal cord (Dum and Strick, 2005). Of the three hand areas, the most posterior one projects directly to the motor neurons in the spinal cord (Rathelot and Strick, 2006). The other two project mainly to interneurons in the spinal cord. This difference might be taken as evidence that the posterior area is more primary in its control of movement. A different explanation, however, may better account for the connectional pattern. The direct projection to the spinal motor neurons, bypassing the spinal interneurons, appears to relate to the control of dextrous manipulation of objects. Animals that are good at dextrous manipulation tend to have this direct projection, and animals that have poor manual dexterity lack the direct projection (Heffner and Masterton, 1975, 1983; see also Bortoff and Strick, 1993; Maier et al., 1997). The data suggest that the direct projection from cortex to spinal motor neurons is not an indication of a lower level in a hierarchy, but instead an indication of the control of a specific kind of action that requires a specific neuronal machinery.

The discovery by Rizzolatti and colleagues of more than one hand representation in the lateral motor cortex, therefore, although of profound importance, did not by itself support the hypothesis of a higher-order premotor area that controls a lower-order primary motor area.

Specialized Properties of F4: Coding Space near the Body

Rizzolatti and colleagues reported that neurons in F4 responded in a distinctive manner to sensory stimuli (Fogassi et al., 1996; Gentilucci et al., 1988; Rizzolatti et al., 1981). Somatosensory responses are commonly found in primary motor cortex, but in F4 a single neuron could have a somatosensory and a visual response. These bimodal neurons responded to a touch on the skin. They were so sensitive that they even responded to a light breath of air on the tips of the hairs. Each neuron responded to touch within a specific area of the body surface, the neuron's tactile receptive field. For some neurons the tactile receptive field was on the face, for example covering one cheek, or the chin, or one half of the forehead, or in some cases the entire face bilaterally. Other neurons had a receptive field on the hand or arm. Rarely a neuron might have a tactile receptive field on the torso. Each bimodal neuron responded not only to a touch on the skin in the receptive field, but also to the sight of an object near or approaching the tactile receptive field. Most neurons responded best when the object was within about 20 cm of the skin, although neurons varied in this spatial extent. These properties suggested that the bimodal neurons encoded not merely the space on the body, but also the visual space immediately surrounding the body, the extrapersonal space.

The bimodal neurons had sophisticated spatial properties. For example, a neuron with a tactile receptive field on the forehead responded to visual stimuli immediately near the forehead. Regardless of where the monkey's eyes were directed, the visual receptive field remained anchored to that same location on the face (Fogassi et al., 1996; Gentilucci et al., 1983). The neuron, therefore, did not respond in relation to the location of the stimulus on the retina; instead it responded in relation to the proximity of the stimulus to a part of the body.

During the 1990s, Charlie Gross and I performed a series of experiments on these bimodal neurons and further elaborated on their properties and distribution in cortex (Graziano et al., 1997a, b; Graziano et al., 1999; Graziano et al., 1994). It is worth noting that they are not necessarily perfectly confined within area F4 as it is usually drawn. As shown in Figure 4-4, our maps of bimodal responses showed a clustering of cortical sites roughly posterior to the bend in the arcuate sulcus, more or less in the location of the dorsal part of F4 (Graziano and Gandhi, 2000). The exact size and location of this cluster varied among monkeys, but the essential pattern was the same across monkeys. We called this area the "polysensory zone" (PZ).

Many of the neurons in PZ were trimodal, responding not only to tactile and visual stimuli but also to auditory stimuli (Graziano et al., 1999). In these cases, the neurons responded to any object within a specific region of space near the body, regardless of the sensory modality through which the monkey detected the object. The neurons also responded to remembered stimuli in the dark (Graziano et al., 1997b). For example, we studied a neuron that responded to objects near the left cheek. An object was shown to the monkey near the right cheek; the lights were turned out; then the monkey's head was rotated such that the unseen object was now near the left cheek. In this condition, the neuron began to respond, as though to the remembered location of the object. The response was not caused by incomplete darkness, or any direct sensory input from the object, because the effect could be obtained even when the object was silently removed in the dark. The neuron did not stop firing until the lights were turned on again and the absence of the object was revealed to the monkey.

These different properties of the multimodal neurons suggest that they monitor the locations of objects near the body through touch, vision, audition, and memory. Why such sensory and cognitive properties should be present in a motor area was unclear. Rizzolatti and colleagues (Gentilucci et al., 1988; Rizzolatti et al., 1981) hypothesized that the multimodal neurons served the general function of the spatial or sensory guidance of movement. In this interpretation, F4 is a higher-order area, a true premotor area, in contrast to the primary motor cortex that controls the implementation details of movement. Initially our interpretation was similar (e.g., Graziano and Gross, 1998).

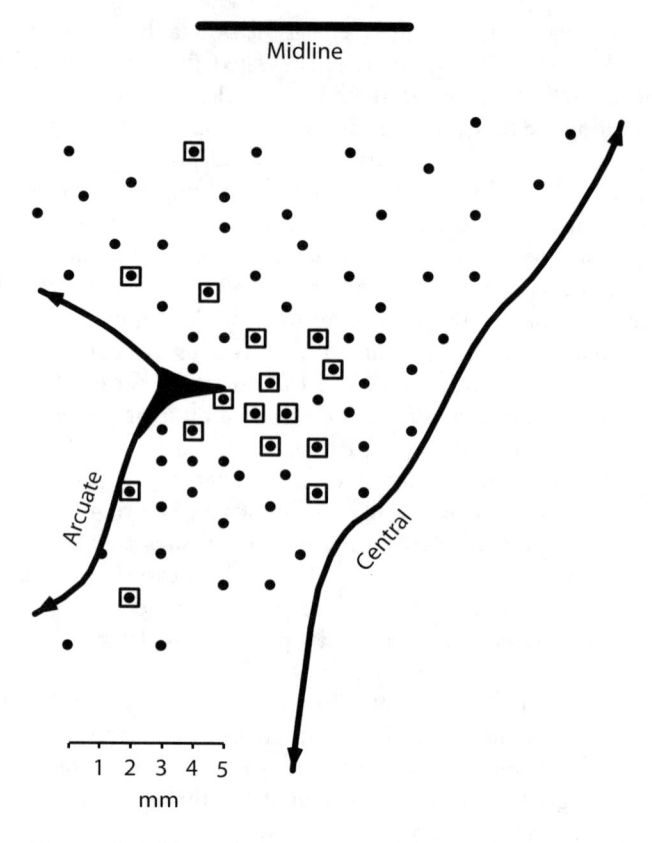

Figure 4-4 The polysensory zone (PZ) in the precentral gyrus plotted in one example monkey. Each dot represents a tested cortical site. Each square represents a site at which polysensory, visual-tactile, or visual-auditory-tactile responses were obtained. PZ is best described as a loose cluster rather than a discrete area with hard borders. Adapted from Graziano and Gandhi (2000).

Our more recent electrical stimulation experiments, however, suggest a different interpretation. F4, or at least the polysensory part of it that we termed "PZ," appears to emphasize a more specific function than the overall sensory or spatial guidance of movement. Stimulation within PZ almost always results in a set of defensive or protective movements (Cooke and Graziano, 2004a, b; Graziano, Taylor, et al., 2002). These movements are fast, reliable, and match in detail the movements that monkeys make when presented with an actual threat such as an air puff to the face or a looming object (Cooke and Graziano, 2003). In our present interpretation, the tactile, visual, and auditory input to PZ is used for the specific function of the defense of the body surface and the maintenance of a margin of safety around the body.

4. Hierarchy in the Cortical Motor System

Rizzolatti and colleagues reported that some neurons in F4 had a tactile receptive field on the lips and responded to visual stimuli near or approaching the lips. In addition to their visual and tactile responses, these neurons also responded when the monkey moved its own hand toward the mouth (Gentilucci et al., 1988). Such neurons seem plausibly suited for guiding feeding movements and are not obviously suited for defensive movements. Why then does stimulation in PZ tend to produce defensive movements and not feeding movements? The reason is probably that the feeding-related neurons and the defense-related neurons are in distinct cortical locations. The F4 neurons with receptive fields around the lips are more ventral and often more anterior to the neurons with receptive fields on other body locations (Graziano et al., 1997a). In our stimulation studies, we consistently found a cortical zone from which defensive movements could be evoked, and an adjacent cortical zone, just ventral and sometimes also anterior to it, from which hand-to-mouth movements could be evoked. One possibility, therefore, is that our hand-to-mouth zone and our defensive zone are contained within F4 as described by Rizzolatti and colleagues.

Specialized Properties of F5: A Library of Grasp Actions

Perhaps the most profound discoveries of Rizzolatti and colleagues relate to area F5. As shown in Figure 4-3, it is located in the ventral, anterior part of the motor strip. Although it is sometimes depicted on the cortical surface, much of F5 is buried in the anterior bank of the arcuate sulcus.

Rizzolatti and colleagues found that neurons in F5 were active when the monkey grasped objects with the hand or the mouth (Murata et al., 1997; Raos et al., 2006; Rizzolatti et al., 1988). Many neurons had a preferred grasp. For example, a neuron might respond most during a precision grip of a small object; another neuron might respond especially well during a whole-hand grip of a large object. A more recent experiment showed that when the chemical *muscimol* was injected into F5, thereby temporarily inhibiting the neurons in the area, the monkey developed a specific inability to grasp objects with the fingers (Fogassi et al., 2001). The hand shape was not well modulated for the specific object to be grasped. Rizzolatti and colleagues hypothesized that the neurons in F5 encoded a library of useful hand and mouth actions.

F5 is at the anterior edge of the area of cortex that projects directly to the spinal cord (He et al., 1993). At least some direct projection from F5 to the spinal cord is likely, although it is probably not a dense projection. This possibility of a spinal projection raises the question of whether F5 is truly hierarchically above the primary motor cortex. One study approached this issue of hierarchy by combining electrical stimulation with chemical inactivation (Shimazu et al., 2004). Stimulation of F5 resulted in measurable neuronal activity in the spinal cord. However, stimulation of F5 immediately after the primary motor cortex was inactivated with injections of muscimol did not result in measurable neuronal activity in the spinal cord. This result suggests that the primary motor cortex is necessary for the normal motor output of F5, supporting the hypothesis that F5 stands at a higher level.

Our own stimulation studies also support the hypothesis that F5 is not at the same hierarchical level as the more posterior motor areas. We found that stimulation in this anterior ventral region, and especially within the posterior bank of the arcuate sulcus, evoked weak, vague movements of the hand that were not as clear or reliable as the movements evoked in the more posterior cortex. At most of these anterior sites no movement was detected unless the current was raised to 200 or 300 microamps, about ten times the thresholds obtained in the more posterior F4.

In our stimulation studies, we consistently found a region of cortex from which hand-to-mouth movements could be evoked at low threshold. As discussed briefly in the last section, this hand-to-mouth zone is generally located just ventral to, or just ventral and anterior to, the polysensory sites in F4. One possibility is that this hand-to-mouth zone corresponds to ventral F4. Another possibility, however, is that it corresponds to the posterior part of F5. The most cautious description of our stimulation experiments with respect to F5 is that we obtained weak and inconsistent movements from anterior F5, especially the portion within the arcuate sulcus, but that the results of stimulating in posterior F5 are not clear because the exact border between F4 and F5 was not certain to us.

Mirror Neurons in F5

Arguably the most influential finding in F5 was the discovery of mirror neurons (Di Pellegrino et al., 1992; Gallese et al., 1996). A mirror neuron has a motor component. The neuron responds when the animal makes a specific movement. For example, a mirror neuron may become active during a precision grip. Unlike a purely motor neuron, the mirror neuron also becomes active when the animal views someone else performing the same act. It has matching sensory and motor response properties, responding during the execution of and viewing of a specific action. The most common interpretation of mirror neurons is that they represent a mechanism by which the animal understands the actions of others through simulating how it would perform the action itself (Gallese et al., 2004; Rizzolatti and Craighero, 2004). In this interpretation, the motor machinery in F5 doubles as perceptual machinery for the comprehension of the acts and goals of others.

Similar mirror response properties have been demonstrated in the human brain in functional imaging experiments (Grafton et al., 1996; Iacoboni et al., 1999). Certain cortical areas are robustly activated by the performance of actions and also by the observation of actions. These cortical areas include parts of premotor cortex, parietal cortex, and superior temporal cortex. These areas have come to be termed the "mirror neuron network" (Rizzolatti and Craighero, 2004).

The hypothesis that mirror neurons participate in the perception of the actions of others is an extension of a much older hypothesis, the motor hypothesis of speech perception (Liberman et al., 1967). In that hypothesis, while listening to another person speaking, we categorize the phonemes by

4. Hierarchy in the Cortical Motor System

simulating with our own motor machinery how we would produce the sound ourselves. The hypothesis of mirror neurons is essentially a generalization from speech to all actions. The hypothesis is that primates understand the actions of others by simulating those actions with the motor machinery.

The mirror neuron story is one of the successes in integrative neuroscience, pulling together social, cognitive, and physiological studies into one framework. Two noteworthy speculations have been added to the mirror neuron story. One is that mirror neurons in area F5 of the monkey represent the evolutionary precursor of Broca's speech area in humans (Rizzolatti and Arbib, 1998). In this speculation, speech is the prime example of a series of acts by one individual that must be understood by a second individual. Hence mirror neurons, designed for perceiving the actions of others, might be used in humans for speech comprehension. In this view, Broca's area is an anterior extension in the human brain of the ventral premotor cortex. Furthermore, although language is usually considered a verbal process, it also involves hand gestures. The monkey F5, with its neurons that are active during hand and mouth movements, may have provided a useful substrate that was expanded through evolution into a processor of language. Note that this fascinating speculation would reinstate the border of human premotor cortex as Campbell originally drew it in 1905, with the ventral premotor cortex extending anteriorly to include Broca's area (see Figure 2-5).

A second speculation is that individuals with autism, lacking the ability to fully understand the goals or social gestures of others, may have been born without proper mirror neurons (Williams et al., 2001). In this view, mirror neurons are so central to the perception of the actions of others, and therefore so central to social interaction, that a lack of mirror neurons should result in a severe social deficit such as is seen in autism. Indeed, when the mirror neuron network is examined in individuals with autism, it is less active than in normals (Dapretto et al., 2006; Iacoboni and Dapretto, 2006). It is not yet known, however, whether the below-normal activity in the mirror network caused the autism, or the other way around.

The main caveat to the mirror neuron story at this time concerns the correlative nature of the experiments. Almost all neurophysiology has become dependant on correlations between brain activity and external states. It has become standard to show that a brain area's activity is correlated with function X, and to optimistically conclude that the brain area must therefore control function X. In the case of mirror neurons, the neurons become active while the animal observes and presumably perceives the actions of others. There is little if any evidence, however, that these neurons actually help cause the perception of the actions of others. If the hypothesis is correct, then a lesion to the mirror neurons in the motor system should result in a loss of ability to correctly comprehend the actions of others. For example, a temporary inhibition of a mirror neuron area in the human brain should cause the person to be measurably less able to understand the actions of other people, while leaving intact other perceptual abilities. Such an experiment has not yet been done. In addition to the lesion approach, another possible causal approach would involve electrical stimulation. Consider an experiment in which a monkey is trained to perceptually categorize a set of actions presented to it on video. If the monkey is observing action A, but electrical stimulation is applied to a cluster of mirror neurons in the motor system that is specific to action B, will the monkey be more likely to make a mistake and miscategorize the action as B? Such an experiment would follow the logic used brilliantly by Newsome and colleagues (Salzman et al., 1990) to study perception in the visual cortex of monkeys. Until causal experiments of this nature are performed, the mirror neuron story remains an exciting hypothesis.

THE DORSAL PREMOTOR CORTEX

Wise and colleagues were the first to systematically study the single-neuron response properties in the dorsal premotor cortex (Weinrich and Wise, 1982; Weinrich et al., 1984; Wise et al., 1983). In their experiments, monkeys were trained by means of juice rewards to make movements in reaction to sensory cues. For example, the illumination of a button might instruct the monkey to reach for the button and press it. In this paradigm, neurons in the dorsal premotor cortex responded with a burst of activity time locked to the sensory stimulus. The distinction between premotor and primary motor was not absolute. Responses to the sensory cue were found in both areas but were more frequently encountered and were more robust in the premotor area. One interpretation is that the dorsal premotor cortex is relatively more concerned with sensory-motor integration, whereas the primary motor cortex is more concerned with motor execution. The effects of lesions to the dorsal premotor cortex in monkeys also suggested that this cortical area may participate in learning arbitrary associations between sensory stimuli and motor responses (Passingham, 1985, 1986).

In some experiments, Wise and colleagues used a delayed movement paradigm (Weinrich et al., 1984). Monkeys were presented with two visual cues in sequence. The first visual cue (the instructional cue) indicated the movement that was to be made, and the second visual cue (the go cue) indicated to the monkey to initiate the movement. Delays up to several seconds could be inserted between the two cues, requiring the monkey to keep in mind a specific planned motor act. In this paradigm, neurons often responded to the initial instructional cue, maintained an elevated firing rate during the delay period as if preparing for the upcoming movement, and then responded with a final burst of activity around the time of the movement execution. Neurons in the primary motor cortex often showed delay activity, but the delay activity was more common and more robust in the dorsal premotor cortex.

The results of Wise and colleagues do not support a categorical distinction between premotor cortex and primary motor cortex. Instead the properties change in a graded fashion from the primary motor cortex, to the PMDc and PMDr. The more rostral cortex shows greater neuronal responses to instructional cues and during delay periods. It has become increasingly common to describe PMDr as a higher-order area that is involved in cognitive processes and is more akin to prefrontal cortex than to motor cortex. PMDr is not densely connected to the spinal cord (He et al., 1993) and is densely connected to the lateral prefrontal cortex (Takada et al., 2004), and its neurons respond in relation to learned task cues in a manner similar to neurons in prefrontal cortex (Brasted and Wise, 2004; Muhammad et al., 2006). In contrast, PMDc seems more akin to motor cortex than to prefrontal cortex. PMDc has a substantial projection to the spinal cord (He et al., 1993), and its neurons are particularly active during reaching movements (e.g., Churchland et al., 2006; Cisek and Kalaska, 2005; Crammond and Kalaska, 1996; Hocherman and Wise, 1991; Johnson et al., 1996; Messier and Kalaska, 2000).

In our experiments, we found a cluster of sites from which stimulation evoked apparent reach-to-grasp movements, including an outward projection of the arm, an opening of the hand as if shaping to grasp, and typically a rotation of the wrist and forearm that oriented the grip away from the body (Graziano et al., 2005). The cluster of sites approximately matched the typical location of PMDc. As described in Chapter 2, this cluster from which apparent reaches can be evoked was first described by Beevor and Horsley (1887). When we stimulated in more anterior cortical regions, presumably in PMDr, we evoked no movement even at high currents.

MEDIAL PREMOTOR AREAS

Penfield and Welch (1951) first described SMA in the monkey and the human brain as a representation of the body on the medial wall of the hemisphere. Woolsey et al. (1952) confirmed SMA in the monkey brain, describing it as a rough somatotopic map with the legs in a posterior location and the face in an anterior location. The representations of different body parts were found to overlap extensively. Stimulation of many sites evoked bilateral movements and sometimes movements of all four limbs. This overlapping somatotopic map in SMA was confirmed by others (Gould et al., 1986; Luppino et al., 1991; Macpherson, Marangoz, et al. 1982; Mitz and Wise, 1987; Muakkassa and Strick, 1979).

Three main hypotheses have been proposed for the function of SMA: coordinating temporal sequences of actions (Gaymard et al., 1990; Gerloff et al., 1997; Jenkins et al., 1994; D. Lee and Quessy, 2003; Mushiake et al., 1990; Roland, Larsen, et al., 1980; Roland, Skinhoj, et al., 1980), bimanual coordination (Brinkman, 1981; Serrien et al., 2002), and the initiation of internally generated as opposed to stimulus driven movement (Halsband et al., 1994; Matsuzaka et al., 1992; Roland, Larsen, et al., 1980; Roland, Skinhoj, et al., 1980). The data, however, tend not to support an exclusive role of SMA in any of these functions. Indeed, SMA is demonstrably active during simple, nonsequential, unimanual, and stimulus-cued movements (Picard and Strick, 2003).

Perhaps one reason for the ambiguity of function in SMA is that it was initially studied as a single motor area, yet it appears to be heterogeneous. At least six areas are now commonly recognized within what was once termed "SMA." The most anterior portion is now commonly termed "pre-SMA" (He et al., 1995; Luppino et al., 1991; Matsuzuka et al., 1992). It has sparse or no connections to the spinal cord or the primary motor cortex and has extensive connectivity with prefrontal areas (Bates and Goldman-Rakic, 1993; Dum and Strick, 1991; He et al., 1995; Lu et al., 1994; Luppino et al., 1993). The SEF is a relatively anterior portion of the SMA that, when stimulated, evokes head and eye movements and perhaps movements of the limbs and torso (Chen and Walton, 2005; Russo and Bruce, 2000; Schlag and Schlag-Rey, 1987; Tehovnik and Lee, 1993). Dum and Strick (1991) hypothesized on the basis of cytoarchitecture and connections to the spinal cord that the portion of SMA in the cingulate sulcus, on the medial part of the hemisphere, can be split into three separate areas, the cingulate motor areas. The functions of the cingulate motor areas have not yet been systematically studied. SMA proper in monkeys has now been confined to a region on the crown of the hemisphere just anterior to the primary motor leg representation. SMA proper projects directly to the spinal cord and therefore may belong to the set of primary output areas of the cortical motor system (e.g., Dum and Strick, 1991; Galea and Darian-Smith, 1994; Macpherson, Wiesendanger, et al., 1982; Murray and Coulter, 1981; Nudo and Masterton, 1990; Toyoshima and Sakai, 1982).

Some hints of the possible functions of SMA can be gleaned from electrical stimulation studies. Some of the movements described by early investigators suggest highly specific actions. Penfield and Welch (1951) in their study of the human SMA stated, "One gains the impression in some cases that the contralateral hand is raised in preparation for its use in a complicated act, and that meanwhile the gaze is directed by movement of the head and eyes toward the hand" (p. 310). They also described "stepping movements," bilateral movements of the extremities, and movements that converged on seemingly meaningful postures.

These suggestive initial observations from stimulation have unfortunately not been pursued as thoroughly as they could have been. Almost all stimulation studies in SMA have focused on somatotopic mapping (e.g., Gould et al., 1986; Luppino et al., 1991; Mitz and Wise, 1987). Although these reports note that complex movements can be evoked from SMA, they place relatively little focus on the possible behavioral meaning or the kinematics of the evoked movements. The use of stimulation to probe function has flourished mainly among researchers who study eye movements. Stimulation of the SEF region of SMA evokes coordinated saccadic eye movements and conjoint head and eye movements that closely match the metrics of normal gaze shifts (Chen and Walton, 2005; Martinez-Trujillo, 2003; Russo and Bruce, 2000; Schlag and Schlag-Rey, 1987; Tehovnik and Lee, 1993).

In our own stimulation studies we explored most of the SMA proper, but not the cingulate motor areas, the SEF, or the pre-SMA. Stimulation within SMA proper evoked complex movements of the limbs and tail, consistent with

4. Hierarchy in the Cortical Motor System

previous reports (Foerster, 1936; Luppino et al., 1991; Penfield and Welch, 1951; Woolsey et al., 1952). Many of the movements resembled climbing or leaping actions. For example, stimulation of one site caused the left foot to press down against the floor of the chair; the right foot to lift and reach forward with the toes shaped as if in preparation to grasp; the left hand to reach toward a lower, lateral position while shaped as if in preparation to grasp; the right hand to reach toward a position above the head while shaped as if in preparation to grasp; and the tail to curl to one side. The macaques in our experiments use their long, stiff tails mainly as balance devices during locomotion, and therefore the tail movements evoked by stimulation of SMA are consistent with a possible role in locomotion. The movements we observed never resembled simple stepping movements. They did not appear to mimic walking on a flat surface. Rather, the movements resembled the highly complex interactions of the body with a challenging, obstacle-strewn environment.

The hypothesis that the medialanterior cortex plays a specific role in locomotion is not new. Tehovnik and Yeomans (1987) found that electrical stimulation in a medial anterior area of the rat cortex evoked forward locomotion and circling, complete with head and whisker movements typical of normal rat locomotion. Whether this cortical area in the rat corresponds to the monkey SMA is not clear. However, the results suggest that some aspects of locomotion may be cortically localized in both species.

INTERPRETING THE PREMOTOR ZOO

The sections above summarize the diverse properties that have been reported for a zoo of premotor areas including at least two ventral premotor areas, two dorsal premotor areas, and as many as six medial premotor areas. The properties described for each area are typically found to some extent in all areas that have been tested. The borders are fuzzy and the differences are of degree rather than of kind, not only among the premotor areas, but also between them and the primary motor cortex. The cortical motor system is unquestionably heterogeneous. Different regions have different functional properties and connections. Yet no systematic or underlying explanation has emerged. The cortical motor system has yielded a long list of disparate observations without a theory.

One possible organization of the cortical motor system is diagrammed in Figure 4-1. Three broad regions of cortex are included in this diagram: A set of parietal areas involved particularly in sensory-motor integration, a set of rostral premotor areas that influence movement control without strong direct projections to the spinal cord, and a mosaic of cortical output areas that organize and control the animal's behavioral repertoire. This overarching hierarchical organization is a nod to tradition. Neuroscientists are perennially seduced by the concept of hierarchy in which simpler areas are connected to more complex areas, often culminating in the prefrontal cortex. To the extent that the present depiction resembles a traditional hierarchy, it should be viewed with great skepticism. Our experiments have focused almost exclusively on the set of cortical output areas, and it is here that the data suggest an organization far from the traditional one.

In our proposal, the highly dimensional motor repertoire is rendered onto the motor output sector of frontal cortex. As a result of this dimensionality reduction, the representation of movement in this large swath of cortex is partially separated into functionally different zones, to the extent that the behavioral repertoire naturally divides into different categories of action. These zones do not resemble true areas. Rather they are graded hills that emerge as a result of statistically common elements in the animal's repertoire. These output zones are approximately at the same hierarchical level, projecting directly to the spinal cord or to subcortical motor nuclei.

Different higher-order functions have been attributed to these different cortical zones. For example, motor preparation signals are common and robust in the PMDc (e.g., Weinrich et al., 1984). Visuospatial information is more prevalent in PMVc, where neurons have clearly defined tactile and visual receptive fields that emphasize the space near the body (Graziano et al., 1997a; Rizzolatti et al., 1981). In the SMA, various studies have suggested a relative emphasis on planning sequences of actions, bimanual coordination, and internal rehearsal of movement (e.g., Brinkman, 1981; Matsuzaka et al., 1992; Mushiake et al., 1990; Roland, Larsen, et al., 1980; Roland, Skinhoj, et al., 1980). Neurons in the primary motor hand area are relatively more correlated with joint rotations and muscle forces, whereas neurons in premotor areas are relatively more correlated with direction in space (Kakei et al., 1999, 2001).

These findings have generally been interpreted in the traditional framework of a set of premotor areas that specialize in different higher-order aspects of movement and a primary motor cortex that implements the details of muscle activations. Yet these differences among cortical zones might be better explained as the result of a rendering of a heterogeneous movement repertoire onto the motor cortex.

For example, stimulation in the PMDc evokes movements that resemble reaching to grasp objects. The action mode of reaching presumably requires short-term planning because it necessarily involves a delay before a goal is reached. It also presumably involves at least some control of external spatial variables to guide the hand to targets.

Stimulation in the primary motor hand area evokes movements that resemble manipulation of objects. Manipulation of objects that are already in grasp, in contrast to reaching, may require less planning, less control of Cartesian spatial location, and more control of muscle forces and individual joint rotations in the fingers, wrist, and forearm.

Defensive reactions are emphasized in the PMVc, specifically in the dorsal part of this region. Defensive movements require processing of sensory events, especially visual and tactile events. They require a processing of spatial locations and trajectories near the body.

Stimulation of the SMA evokes movements that resemble leaping or climbing actions. Complex locomotion that negotiates obstacles in the environment along a desired path might rely especially on an internally generated sequencing of events, and certainly on coordination across the two sides of the body.

Perhaps these kinds of associations between specific action modes in the normal repertoire and styles of processing can explain some of the higher- and lower-order functions relatively emphasized in different cortical zones. The hypothesis is not that planning is solely in the PMDc, that processing of movement sequences is solely in SMA, or that any other function is exclusively linked to one kind of behavior or exclusively controlled in one cortical zone. Rather, each type of behavior requires a broad range of processing styles and a control of many movement variables. Different types of behavior rely on somewhat different mixtures of movement skills, and therefore zones within cortex that come to emphasize different actions have overlapping but nonidentical movement properties.

The hypothesis of a rendering of the highly dimensional movement repertoire onto the two-dimensional cortical surface can explain in great detail the topographic arrangement of the primary motor cortex, caudal premotor areas, SMA, and even FEF and SEF. This can be shown by taking an approximate, multidimensional description of a monkey's movement repertoire, and mathematically flattening it onto a sheet while preserving as much continuity among the movement representations as possible. Such a flattening results in an organization that closely matches the actual organization in the monkey cortex. This method is discussed in greater detail in Chapter 10.

Emergent Hierarchies

To a first-order approximation, in the scheme proposed here, the cortical output zones are on the same hierarchical level, each emphasizing a different behavior and all controlling movement in parallel through direct projections to the spinal cord. Yet to a second approximation, this strictly modular view is not correct. The output zones are laterally interconnected and the lateral information flow is likely to be asymmetric, resulting in subtle emergent hierarchies. For example, consider the hand-to-mouth zone and the manipulation zone. The hand-to-mouth zone controls a type of action that requires integration across the lips, tongue, neck, arm, and hand. The manipulation zone controls a more local type of action that includes a shaping of the hand. The anatomy suggests that both of these zones function at least partly in parallel, sending direct output to the spinal cord. Yet optimum efficiency predicts that the hand-to-mouth-zone should also recruit the manipulation zone, using its circuitry to aid in the control of the hand.

As an analogy, imagine a pair of painters. One is a workman who paints houses, and the other is an artist who paints portraits. They are equally skilled at their assigned tasks. But because of the nature of their skills, the portrait painter could help paint a house whereas the house painter could not help much in painting a portrait. In an ideal use of resources, therefore, the house painter recruits the portrait painter to help in his jobs, the two of them working together; but the portrait painter rarely if ever recruits the house painter.

A subtle hierarchy emerges between two entities whose skills emphasize different tasks. In the same way, the hand-to-mouth zone might recruit the manipulation zone. Hierarchies of this emergent nature should be expected to crop up among the cortical output zones, insofar as these zones emphasize different kinds of actions, are mutually interconnected, and can recruit each other's circuitry to different extents.

This speculation of a subtle emergent hierarchy is consistent with several classical observations. Vogt and Vogt (1919) and Bucy (1933) found that electrical stimulation of the premotor cortex evoked movements through two pathways. One pathway involved lateral connectivity to the primary motor cortex. This pathway could be disrupted by lesions to the primary motor cortex. The second pathway involved a direct descending projection from the premotor cortex. This pathway could be disrupted by cutting the white matter under the premotor cortex. These observations, as relevant now as they were at the time, framed the essential conundrum of the premotor cortex. Does it operate in parallel with the primary motor cortex, in which case the names are misleading and both areas are equally primary? Or does it operate in series with the primary motor cortex, in which case the premotor cortex stands at a higher level than the primary motor cortex? The current speculation potentially solves this conundrum. The cortical output zones operate in parallel, emphasizing different parts of the movement repertoire. However, they also interact laterally, recruiting each other's circuitry in an opportunistic fashion, resulting in subtle emergent hierarchies.

More than One Function for each Cortical Zone

It is extremely difficult to pin a specific function to a brain area. In fact, it is rarely valid to pin a specific function to any part of the body. Consider the function of the human foot. If any part of the body has an unambiguous function, it is surely the foot. It is for locomotion. The major evolutionary pressures that shaped it were related to locomotion, and therefore it is well built for that purpose. But consider the range of purposes to which you put your foot. You might kick soccer balls or other people. You might tap a rhythm. You might step on a bug. You might step on a fallen book to drag it toward yourself. You might use one foot to scratch an itch on the other calf. Your foot may play an essential role in social interaction, if someone tickles it. The foot, though seemingly built for one purpose, is used opportunistically for any purpose that is convenient. It is therefore nonsense to declare that the behavioral purpose of the foot is locomotion, though that may be its most salient purpose. Instead, it is used for a constantly reinvented set of functions, some closely related and others quite idiosyncratic. As a result, evolution may begin to shape the foot toward additional functions, as happened for example in the case of the forefoot of the bat. Evolution does not suffer from functional fixity. The same caution presumably applies to the functions of brain areas.

4. Hierarchy in the Cortical Motor System

Consider the SMA. Stimulation of it in monkeys evokes especially complex movements that sometimes combine all four limbs and that resemble climbing or a shaping of the limbs in preparation for leaping. Do we then suggest that the function of SMA is to control complex locomotion? Such a suggestion seems overly specific, especially because humans have an SMA and yet do relatively little climbing. Instead, we suggest that the control of complex locomotion in an obstacle-strewn environment is one function in a bundle of functions that may be able to coexist in SMA, sharing circuitry and clustering together.

One salient property of climbing is its reliance on bilateral coordination. Bilateral representation of movement can be found throughout the motor cortex, but it is more pronounced in SMA. This bilateral control of the body equips SMA to participate not just in climbing or other complex locomotion, but more generally in any action that relies heavily on bilateral integration. In this speculation, bilateral integration and climbing are two functions that share enough in common to be approximately colocalized, drawing on the same cortical circuitry.

A second salient property of a monkey's climbing is its internally controlled sequencing of events. In reaching-to-grasp, the reach is physically required to come before the grasp. But in climbing or other acts of locomotion through a complex, obstacle-strewn environment, where the supports for the hand and foot are in arbitrary locations, the many movement components do not have any physically required order. The sequence must be internally planned. The slightest mistake in the precision of the sequence could result in a fatal fall. This ability to internally generate temporal sequences of actions need not be limited to climbing or the SMA. It could be recruited for a variety of behaviors. Indeed it may be especially useful in the manipulation of objects, and therefore some degree of sequence planning might be expected within the primary motor hand area as well, as has been reported (Lu and Ashe, 2005).

In this hypothesis it is impossible to decide if SMA is a higher-order area for emphasizing movement sequences and internal rehearsal, or lower order for directly participating in motor output for a particular kind of action. A deeper hypothesis is that functions that are similar or that can profitably share machinery end up trending near each other. The result is a motor cortex that is not divided up into neat areas each with a special function. Rather the cortical motor real estate contains a vast, smeared overlap of functionality with a heterogeneity that is driven by the highly dimensional, statistical structure of the animal's natural behavioral repertoire.

Chapter 5

Neuronal Control of Movement

INTRODUCTION

How do neurons in motor cortex control movement? Two main types of experiment have addressed this question. One type focuses on the descending pathways that map specific points in motor cortex to specific muscles. For example, the activity of a neuron in cortex might affect a set of muscles, activating some and inhibiting others (e.g., Cheney and Fetz, 1985). An underlying assumption in this type of research is that once the mapping from cortex to muscles is discovered, then the functions of motor cortex are largely understood.

A second type of experiment focuses on the activity of single neurons in motor cortex while the animal, usually a monkey, performs a complex task (e.g., Georgopoulos et al., 1986). Aspects of the task, such as the speed or direction of the hand, are correlated with the firing rate of the cortical neurons. The underlying assumption in this type of research is that the relationship between cortex and muscles is more complex than a muscle map. Instead, the firing of a cortical neuron may carry instructions about useful control variables.

These two approaches have resulted in contrasting descriptions of motor cortex. Even within each general approach, researchers have not agreed on the cortical mechanism of motor control. Yet it may be that there is no single correct answer. The answer may depend on the part of motor cortex and the movement under study. A central proposal of this book is that different zones in motor cortex emphasize different modes of behavior that probably have different control requirements. It may be that one type of action, such as manipulation of objects, is more slanted toward muscle or joint control whereas another type of action, such as reaching toward objects, is more slanted toward control of spatial variables.

This chapter reviews these two major approaches to motor cortex. The first part of the chapter summarizes experiments on the direct pathways from motor cortex, through the spinal cord, to the muscles, and how those pathways might control movement. The second part of the chapter summarizes experiments on correlations between the activity of neurons in motor cortex and a variety of control variables related to the arm.

PATHWAYS FROM CORTEX TO MUSCLES

Muscle Force at the Wrist

Evarts (1968) was the first to systematically study the properties of single neurons in the motor cortex of monkeys. In his now-classic experiment, the monkey sat with its arm stabilized in a tube and was trained to use wrist flexions and extensions to apply force to a manipulandum. Because of a system of hanging weights, the direction of force applied by the wrist to the manipulandum could be decoupled from the direction of wrist rotation. Evarts focused his experiment on those neurons in the motor cortex that projected through the pyramidal tract directly to the spinal cord. He identified these pyramidal tract neurons by electrically stimulating the pyramidal tract and measuring the evoked activity in motor cortex. The latency of the evoked activity indicated whether a cortical neuron had had its axon directly stimulated. He then studied these spinally projecting neurons while the monkey performed the wrist task. The neurons were active in a manner monotonically related to the force applied by the wrist. Some neurons were active during force applied in one direction, with greater neuronal activity associated with greater force. Other neurons were active during force applied in the opposite direction.

The results of this experiment are consistent with the view that the neurons in motor cortex, especially those that project directly to the spinal cord, are essentially muscle controllers. The greater the neuronal activity, the greater the signal that passes down the pathway and reaches the muscle, and therefore the greater the torque at the wrist. There are two caveats to this conclusion, however. First, the study is limited to the control of the wrist. Fine control of the wrist and fingers may have evolved a specialized machinery. In primates that manipulate objects with a high degree of skill, the motor cortex projects directly to the spinal motoneurons that control the hand (Bortoff and Strick, 1993; Heffner and Masterton, 1975, 1983; Maier et al., 1997). The control of other body parts, such as the upper arm, involves mainly projections from the motor cortex to spinal interneurons. The direct cortical control of wrist muscles, implied by the Evarts result, therefore might not be directly applicable to other body parts.

A second caveat concerning the Evarts study is that the limb was restrained in a consistent, relatively unchanging state. Kakei et al. (1999) found that if the arm is placed in different configurations, the mapping from cortical neurons to wrist muscles can change. A neuron that correlates with wrist extension, when the arm is in one orientation, may switch and correlate with wrist flexion when the arm is placed in another orientation. About half of the neurons in primary motor cortex showed some change in mapping to the wrist muscles caused by a change in arm configuration.

These findings on neuronal activity during wrist movement are consistent with the view of feedback remapping (Graziano, 2006). In that view, the mapping from cortex to muscles depends on a rich circuitry that includes the motor cortex, the spinal cord, and probably other structures. The state of this

5. Neuronal Control of Movement

circuitry can change depending on signals about the state of the periphery. If the feedback is constant, such as when the limb is maintained in a relatively fixed position, then the circuitry may remain more or less in one state and provide what appears to be a fixed mapping from cortical neurons to muscles. If the feedback from the periphery is changed, such as by putting the limb in a new configuration, then the circuitry is put into a different state and the apparent mapping from cortical neurons to muscles changes.

Spike-Triggered Averaging and Stimulus-Triggered Averaging

Cheney and colleagues (Cheney and Fetz, 1985; Cheney et al., 1985) examined in detail the pathways from single neurons in motor cortex to the muscles of the hand and arm. Their method of spike triggered averaging was briefly discussed in Chapter 3. In these experiments on the monkey motor cortex, the activity of a neuron in motor cortex was measured. At the same time, the activity of a set of muscles in the arm was measured. When the neuron in motor cortex fired a spike, after a latency of approximately 5 to 10 ms, a minute change occurred to the muscle activity (see Figure 5-1). This effect was so subtle that it could be demonstrated only by averaging the results of thousands, in some cases tens of thousands of neuronal spikes. Because of the short latency, the results strongly suggested that the neuronal signal measured in cortex must have passed down the most direct path possible, from the cortical neuron to the motor neurons in the spinal cord, and from there to the muscle. A variant of the technique, stimulus-triggered averaging, used pulses of electrical stimulation applied to the cortex rather than the measurement of

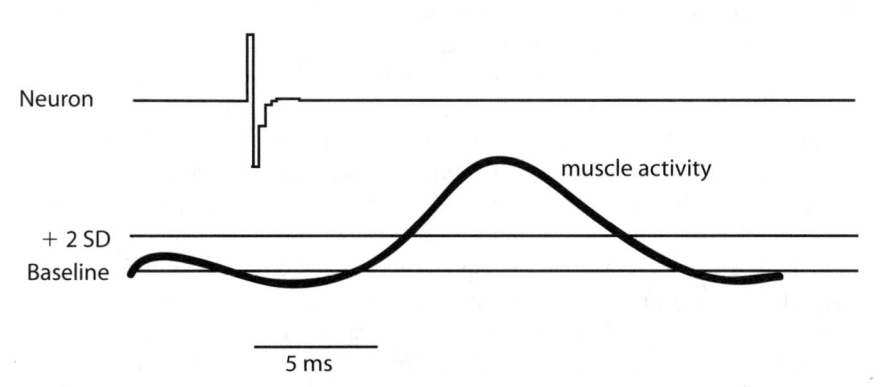

Figure 5-1 Spike-triggered averaging. In this technique the spiking of a neuron in motor cortex is measured. At the same time, the activity in a muscle of the arm is measured. Approximately 5 ms to 10 ms after the neuronal spike, a rise in muscle activity can be observed, implying that the activity of the cortical neuron caused the activity of the muscle. Typically data from 500 to 10,000 neuronal spikes are averaged to obtain this result. The muscle activity trace shown is a schematized illustration of the typical traces found in studies such as Cheney and Fetz (1985).

neuronal spikes in cortex. The causal method of stimulus triggered averaging and the correlational method of spike triggered averaging yielded similar results (Cheney and Fetz, 1985).

These techniques indicated that a neuron or a stimulation point in motor cortex maps to a set of muscles, exciting some muscles and inhibiting others. The finding is consistent with the view of a many-to-many mapping from cortex to muscles. It has helped to support the hypothesis that the motor cortex controls movement via a fixed relationship to muscles wired through a relatively direct descending pathway. Yet this view of a fixed mapping from cortex to muscles is incomplete. The experiments were simplified such that changes in the state of the limb were not considered. When sensory feedback indicates a change of conditions in the periphery, will the mapping from cortex to muscles also change?

Feedback Remapping

A study by Sanes et al. (1992) on the rat motor cortex suggested that indeed the mapping from the primary motor cortex to the muscles can change depending on the state of the limb. These experimenters stimulated the motor cortex of anesthetized rats to map the representation of the front leg. They determined which areas of cortex, when stimulated, evoked activity in the muscles of the limb. They tested first with the limb in a flexed posture, and then with the limb in an extended posture, and found that the map in cortex changed substantially across these two conditions. This study may be the first to show that feedback about a change in limb configuration could alter the specific mapping from the cortex to the muscles.

A study by Lemon et al. (1995) on the human motor cortex also suggested that the mapping from cortex to muscles can change depending on the state of the limb. In this study, the experimenters measured the connection strength between motor cortex and various muscles of the arm and hand. Their method was to activate the primary motor cortex with pulses of transcranial magnetic stimulation (a magnetic method of stimulating the brain through the skull) and to measure the evoked activity in limb muscles. A relatively larger amount of evoked activity indicated a stronger connection from cortex to muscles, and a relatively smaller amount of evoked activity indicated a weaker connection. At the same time, the participant performed a reaching and grasping task. The connection strength between motor cortex and the muscles changed markedly in different phases of the task, suggesting that the mapping from the cortex to the muscles was not fixed but instead was modulated continuously as the reaching task unfolded.

A recent study of ours on the monkey motor cortex indicated that the mapping from primary motor cortex to the biceps and triceps muscles of the arm can change depending on the angle of the elbow joint (Graziano, Patel, et al., 2004). We used the technique of stimulus triggered averaging, stimulating a site in the motor cortex in an anesthetized monkey with pulses of current and measuring the short latency changes in muscle activity. Stimulation of

5. Neuronal Control of Movement

each site in cortex evoked a change in muscle activity within 5 to 7 ms, indicating that the most direct, descending pathway was recruited. Whether stimulation of a site evoked activity primarily in the biceps, primarily in the triceps, or in a relative mixture of the two, depended on the angle at which the elbow was fixed. Some cortical sites could even be switched from mapping mainly to the biceps to mapping mainly to the triceps. Feedback about the angle of the elbow joint therefore modulated the pathway from the stimulated site in cortex, through the spinal cord, to the muscles, causing a change in the mapping from cortex to muscles. These experiments are described in more detail in Chapter 7.

Perhaps the central lesson in the research on descending pathways from cortex to muscles is that the term *pathway* is not quite the correct designation. Activity in motor cortex neurons is not merely transmitted downward to muscles along wires. Instead a rich network intervenes. This network is modulated by feedback from the periphery that influences the spinal and cortical circuitry. When the feedback is held more or less constant, then the circuitry is held in more or less one state, and each neuron in cortex appears to map through that circuitry to a fixed set of muscles. When the state of the periphery varies, the feedback modulates the circuitry, and therefore the mapping from each neuron in cortex through that circuitry to the muscles also changes.

The caveat of feedback remapping is that it is not yet clear just how extensively feedback can alter the mapping from cortex to muscles. The experiments described above focused on feedback about static arm position and demonstrated some degree of change in the mapping from sites in cortex to muscles. How speed, tension, skin pressure, visual feedback from the arm, or other feedback signals might or might not alter the mapping from cortex to muscles remains untested.

The usefulness of a feedback-dependant mapping from cortex to muscles is that it can in principle allow neurons in motor cortex to control a diversity of movement variables, such as direction, speed, hand position, or posture that transcend a fixed pattern of muscle activation. If the network receives feedback information about a specific movement variable, then it can learn to control that variable. We suggested that feedback remapping is the general class of mechanism that allows motor cortex to control useful, complex behavior (Graziano, 2006; Graziano, Patel, et al., 2004). This control of movement variables by motor cortex neurons is discussed in greater detail in the following sections.

SINGLE NEURONS IN MOTOR CORTEX AND THEIR CONTRIBUTIONS TO MOVEMENT

Ever since the groundbreaking experiments of Hubel and Wiesel (1962) on visual cortex, the study of single cortical neurons has become traditional in all sensory and motor systems. For the past 30 years, the study of motor cortex in particular has focused on the properties of single neurons. The technique, however, works somewhat better for studying sensory systems than for studying motor systems. In the visual system, an experimenter can present a visual stimulus and measure the neuronal activity caused by that stimulus. The method is therefore causal. Hundreds of visual stimuli can be tested, to winnow down the exact features of the visual world that trigger activity in the neuron. In the motor system, the experimenter trains a monkey to perform a small set of movements, perhaps eight or ten in total. As the monkey performs these movements, the experimenter measures the activity of neurons in the motor cortex and hopes that this neuronal activity is correlated with some aspect of the ongoing movement. The method is therefore correlational rather than causal, depending on correlations between two things not directly under the control of the experimenter. Moreover, the search space of movements is severely limited by the difficulty and time involved in training the monkey on each movement to be tested. As a result it is much more difficult to obtain a clear answer from motor cortex neurons than from neurons in the visual system. After 30 years, the study of motor cortex neurons is still bogged down in the ambiguities of correlations and alternate interpretations.

Yet a great deal has been learned, especially about general principles of the cortical control of movement. A complete description of the many experiments, arguments, and counterarguments is outside the scope of this chapter. The following sections discuss some of the main experimental results and their possible implications. The relationship between single-neuron properties and the effects of stimulation is further discussed in Chapter 8.

A Population Code Based on Fragments of Behavior

The experiments described above suggest that the connectivity from cortical neurons to muscles is flexible. A neuron in motor cortex does not necessarily contribute to a fixed set of muscle activations. In principle, the firing of a cortical neuron could translate into a complex fragment of behavior that transcends a fixed set of muscle activations.

For example, a movement of the hand to the mouth uses different muscle activations depending on the initial configuration of the arm. The firing of a neuron in cortex could in principle contribute to one set of muscle activations when the arm is in one posture, and contribute to a different set of muscle activations when the arm is in a different posture, thereby helping to move the hand to the mouth from a range of initial conditions. In this hypothetical example, feedback information about the position of the arm alters the network that maps a cortical neuron to the arm muscles, thereby effectively constructing a "hand-to-mouth" neuron.

This hypothetical example makes the point that in a network with feedback, the level of control can be much more complex than a simple muscle map. It is possible to wire up neurons to control complex postures, grip force, hand speed, hand direction, or any other movement variable for which feedback is available. Indeed a neuron need not be limited to controlling one movement variable. In a defensive blocking movement of the arm, for example,

5. Neuronal Control of Movement

the speed, direction, final posture, and final stiffness of the arm are all of importance. A cortical neuron could, in principle, instruct the limb to perform this complex fragment of behavior.

By combining insights from previous single-neuron studies and from our recent electrical stimulation experiments, it is possible to construct a tentative summary of the role of motor cortex neurons in movement, encapsulated in the following three principles.

1. Population code: The activity of a single cortical neuron does not normally cause a movement. Instead, populations of neurons combine their effects, each neuron adding incrementally to the total (Georgopoulos et al., 1982, 1986, 1988).
2. Single-neuron specificity: The incremental contribution that each neuron makes to the population is that it "votes" for a specific movement, its "preferred" movement (Georgopoulos et al., 1982, 1986, 1988). When the neuron is active, it nudges the system toward producing that movement. The greater the activity of the neuron, the greater the nudge it provides toward its preferred movement. The actual movement that the animal produces is therefore a compromise or a weighted sum of the many different preferred movements of the many cortical neurons that are simultaneously active.
3. Neuronal control of complex fragments of behavior: In general, the preferred movement of a cortical neuron is a quirky, complex action learned by the network from the statistics of normal behavior. Activity of a cortical neuron, therefore, incrementally biases the system toward producing a fragment of behavior. Different regions of motor cortex tend to specialize in different parts of the behavioral repertoire. Therefore neurons in different cortical regions have somewhat different properties.

Of these three principles, the first two are not controversial. The third is an extrapolation from our own electrical stimulation experiments and does not represent a consensus view. Indeed there is no consensus view. Although it is generally accepted that each cortical neuron has a preferred movement, the exact nature of those preferred movements remains widely debated. The following sections outline some of the competing views and discuss why we proposed the specific hypothesis that neurons in motor cortex prefer fragments of normal, complex behavior.

Direction Tuning

One of the most influential findings on motor cortex function was the discovery of direction tuning by Georgopoulos and colleagues (Georgopoulos et al., 1982, 1986, 1988; Schwartz et al., 1988). A monkey was trained to move its hand from a central starting location to any one of eight surrounding target locations arranged approximately 12 cm away from the start point (Figure 5-2). At the same time, the activity of neurons in motor cortex was measured. Most neurons

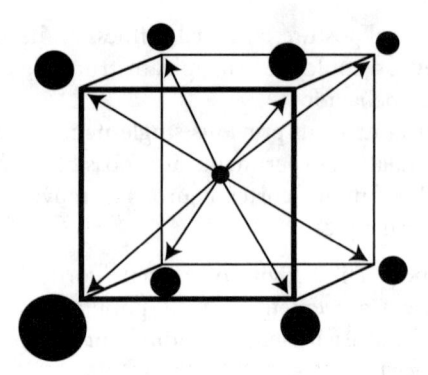

Figure 5-2 Direction tuning of a motor cortex neuron similar to that described in Georgopoulos et al. (1986). A monkey was trained to make hand movements from a central location to eight possible surrounding locations forming the vertices of an imaginary cube. Each reach trajectory was 12.5 cm long. Many neurons in motor cortex were broadly tuned to the direction of the reach, firing more during one direction and less during neighboring directions. In this figure the size of each black dot represents the firing rate of a hypothetical motor cortex neuron during each direction of reach. This neuron prefers a lower, left direction of reach.

in the arm representation in motor cortex were active during this task, their firing rate rising just before the reach and generally remaining high during the reach. A typical neuronal response is shown schematically in Figure 5-2. Here the size of each black circle represents the activity of the neuron as the hand moved toward the target. This example neuron was most active during movements to the lower, near, left target, and less active during movements to the other targets. This response profile is typical in that the neuron fired most during one direction of reach, and fired less well to neighboring directions. In the commonly accepted terminology, the neuron was "tuned" to a "preferred" direction.

Not all neurons had a smooth tuning function. Some neurons preferred two nonadjacent directions, and even those with one tuning peak did not always precisely match the tuning function that was used to fit the data. However, the description of a tuning curve with a single peak was accurate to a first approximation for most neurons.

Georgopoulos et al. (1986) proposed a brilliant interpretation of these findings inspired by the rules of vector algebra. In their interpretation, each neuron is responsible for or "votes" for a specific direction of hand movement. When the neuron becomes active, it incrementally nudges the hand in the neuron's preferred direction. The greater the neuron's firing rate, the more the hand is impelled in that direction. Normally, many neurons are simultaneously active, each neuron biasing the hand in a different direction. These many conflicting directions of movement sum approximately linearly, resulting in a

5. Neuronal Control of Movement

single coherent movement vector for the hand. For example, if a leftward-preferring neuron is twice as active as a rightward-preferring neuron, then at least with regards to those two neurons, the hand will be pushed toward the left. If an upward-preferring neuron and a rightward-preferring neuron are equally active, their contributions will push the hand in a diagonal, upward-and-right direction. During a hand movement, tens of thousands of neurons across the motor cortex are active, some neurons more active than others, each neuron voting for its own preferred hand direction. The overall balance of activity in this population of neurons determines the actual direction of the hand.

This hypothesis of a population of neurons controlling movement has one major caveat. It relates exclusively to the direction of the hand through space. Presumably motor cortex neurons control other aspects of movement in addition to hand direction. Indeed, a neuron might appear to control hand direction when it actually controls something else that happens to correlate with hand direction. These issues were probed in further experiments, described below.

Extrinsic Versus Intrinsic Coordinates

The hypothesis of direction tuning was challenged by Scott and Kalaska (1995, 1997). They trained monkeys to perform a center-out reaching task and plotted directional tuning curves as in Figure 5-2. They found that when the hand direction was held constant, but the joint configuration of the arm was changed by raising or lowering the elbow, neurons in motor cortex changed their activity. The neuronal activity was therefore not exclusively tied to the direction of the hand through space but reflected something else about the details of the arm movement, perhaps the angles of joints or the contraction of individual muscles, something that changed when the elbow was raised or lowered. As a result of this now-classic experiment, a debate emerged over whether motor cortex neurons control high-level spatial variables such as hand direction (termed "extrinsic coordinates") or low-level variables such as joint angles and muscle forces (termed "intrinsic coordinates"). At least one experiment demonstrated convincingly that extrinsic and intrinsic coordinates are both represented in motor cortex (Kakei et al., 1999).

Tuning to Many Movement Variables

Over the past twenty years, dozens of experiments have been performed in an attempt to pinpoint the movement variables that are correlated with the activity of motor cortex neurons.

Neuronal activity in motor cortex is highly correlated with the activity of muscles in the arm, suggesting some degree of direct control of muscles (e.g., Holdefer and Miller, 2002; Townsend et al., 2006). Neurons in motor cortex are sensitive to the starting position of the hand during a reach (Caminiti et al., 1990; Sergio and Kalaska, 2003). They are tuned to the direction of force applied by the hand to a handle (e.g., Georgopoulos et al., 1992; Sergio and Kalaska, 2003). They are tuned to hand velocity during a reach (Moran and Schwartz, 1999; Paninski et al., 2004; Reina et al., 2001). When tested in a different manner, however, the neurons are not consistently tuned to velocity (Churchland and Shenoy, 2007). Motor cortex neurons are not well tuned to the distance traveled by the hand during a reach (Fu et al., 1993). They are also not well tuned to the spatial end point of a hand movement, when tested over a small region of space (e.g., Caminiti et al., 1990; Kettner et al., 1988). When tested over a large region of space, however, motor cortex neurons are well tuned to the spatial end point of a hand movement, and even better tuned to the multijoint configuration of the arm (Aflalo and Graziano, 2007).

It is difficult to interpret this mixture of strong and weak correlations. One difficulty is that different variables are often tested in different ways, or over different ranges. Direction tuning is typically tested over its full 360-degree range, whereas tuning to position is typically tested over a small or arbitrary segment of the working range. Perhaps not surprisingly, therefore, direction tuning is often found to be robust and position tuning is often found to be weak. In addition, experiments sometimes set one type of variable against another, attempting to determine, for example, whether cortical neurons control direction or position, or whether cortical neurons control extrinsic or intrinsic coordinates. The impression to emerge from the many experiments, however, is that neurons in motor cortex do not control one type of variable but instead control mixtures of all variables relevant to normal movement. In our hypothesis, this mixture of variables reflects the tuning of neurons to meaningful fragments of behavior.

Normal Behavior Requires the Control of Combinations of Variables

Consider one of the most common categories of action in the primate repertoire, bringing the hand to the mouth and maintaining the hand near the mouth while the grip interacts with objects in the mouth. To a first approximation, the behavior involves the acquisition of and maintenance of a specific complex joint configuration. This joint configuration includes the elbow flexed and in lower space, the forearm supinated such that the palm faces the mouth, the wrist slightly flexed, and the fingers in a grip posture. The joint configuration is an essential part of the behavior.

In addition to this underlying joint configuration, the behavior involves continuous adjustments to the joint angles to move the hand about the space near the mouth. Muscle force in the fingers is adjusted to regulate the grip strength. Muscle force in the arm and shoulder allows the hand to twist, pull at, or break objects in the mouth. Hand speed is regulated for movements toward the mouth and for smaller movements near the mouth. Hand-to-mouth interaction, therefore, is not a cursor-like translation of the hand from point A to point B but instead requires detailed control of the position, speed, and spatial orientation of the hand, the joint configuration of the arm, and

5. Neuronal Control of Movement

the muscle forces applied by the arm and hand. There is no single variable that fully describes this movement. It involves control of extrinsic and intrinsic coordinates. One salient property of the behavior is that it is clustered around an underlying, canonical posture to which the joints move and about which the joints are continuously adjusted.

As a second example of a common behavior, consider a monkey that sits on its haunches, one hand resting on the ground at its side and the other hand holding a piece of fruit in central space just in front of the chest. The animal rotates the fruit and inspects it. The movements and rotations of the hand are variations on a larger stabilizing posture. The upper arm is vertically downward such that the elbow is at the monkey's side, the elbow is flexed to about 90 degrees, the arm is internally rotated such that the hand is in central space, the wrist is partially flexed, the forearm is supinated such that the grip is aimed at the face, and the grip is closed. On that larger posture, small adjustments are made to all the joints in the arm to move or rotate the object, and to the grip that applies pressure to the object. Again, there is no single variable that describes this movement. It involves a mixture of extrinsic and intrinsic coordinates. A salient property of the behavior is the underlying, canonical posture to which the joints move and about which the joints are continuously adjusted.

Almost all behaviors that use the arm and hand are similar in that they require the control of a range of movement variables, including extrinsic and intrinsic, and they depend to some degree on adjustments around an underlying canonical posture.

Exactly these types of actions are generated when electrical stimulation is applied to sites in the motor cortex of monkeys. We suggest therefore that motor cortex neurons are wired up to the motor network to produce fragments of the normal movement repertoire. If neurons in motor cortex are tuned to fragments of behavior, then two predictions can be made on the basis of the statistics of normal behavior. First, most neurons should be tuned to the joint configuration of the arm, broadly preferring a canonical posture. Second, superimposed on that broad tuning to the canonical posture, each neuron should be tuned to a diversity of other variables relevant to normal movement, such as hand direction and speed. A recent experiment appeared to confirm both predictions (Aflalo and Graziano, 2007). This experiment is discussed in greater detail in Chapter 8.

The Generalized Population Code

The population hypothesis of Georgopoulos et al. (1986), described above, was developed to account for the control of hand direction during reaching. The hypothesis, however, need not be limited to hand direction. It can be generalized to the control of any complex behavior. Consider a set of neurons each of which is tuned to some fragment of complex behavior. In the population hypothesis, when a neuron becomes active, it incrementally nudges the motor output toward the neuron's complex preferred movement. The greater the neuron's activity, the more the system will be impelled toward producing that specific fragment of behavior. Normally many neurons are simultaneously active, each one biasing the output toward a different movement. These many movements sum roughly linearly, resulting in a single movement output.

In this more general formulation of the hypothesis, the basis set of movements on which the population code rests is not a set of hand directions, but instead a set of action fragments that span the normal repertoire. In this general population coding hypothesis, however, it is not clear exactly how the preferred movements of neurons are summed to produce the resultant movement. In a direction tuning experiment, one can sum the directional vectors. The mathematical sum is well defined. But if neurons are tuned to complex and arbitrary movements, rather than to hand direction, then what aspect of movement is summed? Evidence now exists for a linear summation of motor cortex outputs at the level of muscle activity. In a recent study in the cat motor cortex, Ethier et al. (2006) electrically stimulated sites in motor cortex while measuring the activity of limb muscles. They compared the effect of stimulating two cortical sites simultaneously or individually. When the sites were stimulated simultaneously, the muscle output was approximately a linear sum of the outputs obtained from stimulating the sites individually.

This elegant experiment of Ethier et al. (2006) supplies a framework for the most general hypothesis of a population code. Even if the preferred movements of the cortical neurons are arbitrary and complex fragments of behavior, they can sum at the muscle output and thereby contribute to a population average.

A population code of this type, depending on a basis set of fragments of normal behavior, is useful for a variety of reasons. First, it is highly flexible because it allows for linear combinations of a vast number of behaviorally useful movements preferred by different neurons. Second, it is robust. Even if the neurons that are specialized for one particular type of action are damaged, the remaining basis set of movements can produce the action to at least some level of expertise. Third, the population code is optimized for the particular movements useful to the animal. There is no point in a movement basis set that emphasizes mainly reach directions, if pure translations of the hand through space comprise only a fraction of the movement repertoire. Instead, in the present hypothesis, the most common pieces of the repertoire, learned by the network, form the basis set for movement.

A central finding of our stimulation studies is that different behaviorally relevant actions tend to be evoked from different regions of motor cortex. This regional specialization for actions should not be interpreted in too rigid a manner. The fundamental code, in this hypothesis, is a population one. Even if a region of cortex contained neurons all tuned to action A, these neurons would by hypothesis contribute to any movement to the extent that the movement resembles action A. Thus the neurons would be active to some extent,

5. Neuronal Control of Movement

and contribute incrementally, to actions B, C, and so on. One would not expect cortical area A to light up exclusively when the animal performs action A, and cortical area B to light up exclusively when the animal performs action B. Instead, one would expect a pattern of activity distributed across the entire motor cortex with a relative hump in the corresponding cortical location, at least to the extent that the action performed by the animal resembles one of the actions that has special representation in cortex. Likewise, one would not expect lesions of cortical area A to eliminate performance of action A. Rather, one would expect lesions of cortical area A to cause a degradation in performance of action A, as other cortical areas, less well tuned to the action, combine their movement bases to produce the action. However, on the basis of this hypothesis, one does expect that direct stimulation of cortical area A should evoke an approximation to action A, direct stimulation of cortical area B should evoke an approximation to action B, and so on.

Recently, Brecht et al. (2004) reported that artificial activation of a single neuron in the rat motor cortex can cause coordinated, oscillatory whisker movements typical of normal behavior. On the one hand, this result supports the view that neurons in motor cortex are wired up to cause complex actions that are behaviorally relevant. On the other hand, the result seems to be a remarkable departure from the population view of motor cortex. How is it that a single neuron, when active, can cause a movement? In fact the result highlights the subtlety of the population hypothesis. Presumably, the one neuron that was stimulated addressed the larger network, recruiting neurons in cortex and in subcortical nuclei. It is not actually a single neuron, by itself, causing a movement. It is a single neuron addressing a population of neurons, and the population causing a movement. Although one can study the function of a single neuron, its function is defined by its relationship to the larger population.

Chapter 6

What Can Be Learned from Electrical Stimulation?

INTRODUCTION

The first known description of electricity applied to the body for medical reasons is in a treatise from 46 AD by Scriobonius Largus. This Roman physician suggested placing a live torpedo fish on the head to treat headache (Kellaway, 1946). Presumably the powerful electric discharge from the fish stimulated not only the aching portion of the head but also the underlying brain. Arguably, therefore, this application is the earliest example of direct brain stimulation. The use of electricity on the body continued unsystematically for centuries, until Galvani's groundbreaking experiments in 1791. Galvani's studies on frogs, demonstrating that electricity could activate nerves and cause muscle contractions, mark the beginning of modern neurophysiology. As described in Chapter 2, Fritsch and Hitzig (1870/1960) used the technique for the first time to probe the function of specific areas of the cortex. After Fritsch and Hitzig, the technique continued to be refined from a blunt stimulation of the surface of the brain to a targeted application of current pulses through a microelectrode. Microstimulation, as it is often termed, is now one of the principle methods used to study brain function. The following sections provide a brief history and review of the stimulation technique. Hopefully by the end of the chapter the reader will be convinced that the technique, like any other, has its limits, interpretational ambiguities, and uses.

SURFACE STIMULATION

The first systematic use of electricity to study the cerebral cortex dates to Fritsch and Hitzig (1870/1960) who, as detailed in Chapter 2, electrically stimulated the surface of the dog brain and discovered the motor cortex. For nearly a century after Fritsch and Hitzig, surface stimulation continued to be used to map the cortical motor areas, culminating in the work of Penfield and Boldrey (1937) who published a definitive map of surface stimulation of the human motor cortex and Woolsey et al. (1952) who published a corresponding map of the monkey motor cortex.

Penfield's experiments using surface stimulation in the human brain extended beyond the motor cortex. He also studied the somatosensory cortex, auditory cortex, visual cortex, and cortical association areas. Among his most famous studies are those in the temporal lobe in an area that he termed the "interpretive cortex," in which stimulation of specific points appeared to trigger remembered experiences (Penfield, 1959). One patient thought she was hearing a favorite tune played on a phonograph. When the cortical stimulation stopped, she thought the doctor had turned off the phonograph. In reporting these complex effects of stimulation, Penfield grappled with the issue of localization of function. Clearly a specific effect was evoked by stimulation of a specific site. Yet the directly stimulated site could hardly be responsible for the entire effect. Rather, stimulation must somehow activate the circuitry to which that cortical site was connected. One could say that the function of the cortical site was defined by its unknown and possibly widespread connectivity.

HYPOTHALAMIC STIMULATION

Deep stimulation, or stimulation by means of an electrode that was insulated except at the tip and inserted into the brain, began to be used extensively in the 1950s and 1960s. Olds discovered that stimulation of deep structures in the septal region of the rat brain caused an apparent reward (Olds and Milner, 1954). The rat tended to revisit the locations in its cage where it had received brain stimulation. Olds then showed that stimulation in many brain areas especially in the hypothalamus had apparent reward value. The electric reward could be used as effectively as food reward to train bar pressing and maze running.

Further experiments on the hypothalamus revealed that a gamut of emotional and motivational states was on tap, accessible through an electrode. Fearful, aggressive, sexual, predatory, and appetitive behaviors could be evoked by appropriately placed electrodes (e.g., Caggiula and Hoebel, 1966; Hess, 1957; Hoebel, 1969; King and Hoebel, 1968). For example, stimulation of the lateral hypothalamus evoked eating, exactly as if the animal was experiencing hunger (Hoebel, 1969). As long as the stimulating current was applied, for seconds or minutes at a time, the animal ate. As soon as the stimulation stopped, the animal dropped the food as if the hunger state had disappeared. Likewise, stimulation in a different region of the hypothalamus evoked aggression. Stimulation of that location in a rat caused it to attack and kill a mouse in the same cage (King and Hoebel, 1968).

In these experiments the stimulation consisted of a high-frequency train of electric pulses. The behavioral effects of stimulation were robust across a wide range of stimulation parameters. The frequency of stimulation pulses ranged anywhere from 50 to 500 Hz. The currents ranged from 10 to 500 microamps. A train of square wave pulses worked better than a sinusoid, in the sense that a behavior could be evoked with lower current, probably because the sharp onset of each pulse was more effective at discharging neurons. Biphasic pulses with a negative deflection followed by a positive deflection worked better than biphasic pulses of the opposite order, probably because a negative electrical pulse in the extracellular matrix more effectively drove the nearby neurons to their firing threshold. Monopolar pulses tended to damage the brain through a buildup of charge, whereas biphasic pulses appeared not to result in measurable damage even at high currents (e.g., 500 microamps) and long durations (minutes). The parameters therefore could be adjusted and optimized in a number of ways, yet the results were substantially the same. Activation of specific loci in the hypothalamus evoked specific motivated behaviors regardless of the details of the stimulation protocol.

Two fundamental questions can be asked about the technique. First, when a train of electrical pulses is passed through an electrode, what neural elements near the tip of the electrode are directly activated? This issue has been reviewed previously (Ranck, 1974; Tehovnik, 1996; Tehovnik et al., 2006). The answer seems to be that stimulation directly activates a shell of tissue around the electrode tip. The inner and outer diameters of this shell depend on the stimulation parameters. The most excitable elements are the axon hillock and nodes of Ranvier of pyramidal cells or other large projecting cells. To a first approximation, therefore, electrical stimulation generates action potentials within the local tissue around the electrode, though not all elements are equally excited.

The second fundamental question is less about the physics of electrical stimulation and more about the dynamics of networks. Once the local neural tissue is directly stimulated by the electrode, how does the signal spread through connected networks? The question could also be put as follows: when artificial signal is injected into location A in the brain, and that local signal then ramifies through networks, does this spread of signal mimic the contribution that location A normally makes during behavior? An exact match seems unlikely. There are too many reasons for the stimulation-evoked signals to deviate from natural signals, including the unnaturally square time course of the stimulation, the antidromic activation of neurons, and the synchronized coactivation of neurons near the electrode that might not normally be coactive. Stimulation should not be expected to exactly mimic normal function. The results of stimulation suggest, however, that it mimics function well enough to provide useful insight. It successfully identified the basic functions of the main components of the hypothalamus.

EYE MOVEMENT STUDIES

The experiments on the hypothalamus, described above, required the development of fine electrodes that were insulated except at the tip, inserted into the brain, and used to deliver small and precisely controlled currents. Robinson and colleagues borrowed this microstimulation technique to study the control of eye movement in the monkey brain (Robinson, 1972; Robinson and Fuchs, 1969). Stimulation of a site in the superior colliculus resulted in a fast, saccadic movement of the monkey's eyes. The movement closely resembled a naturally occurring saccade in its spatial and temporal dynamics. A map was established in the superior colliculus in which stimulation of different sites evoked eye movements of different directions and distances (Robinson, 1972; Schiller and Stryker, 1972). A similar map, though less finely organized, was found in the frontal eye field (Bruce et al., 1985; Robinson and Fuchs, 1969). Stimulation-evoked eye movements were then studied in several other cortical areas including the lateral intraparietal area and the SEF (Kurylo and Skavenski, 1991; Schlag and Schlag-Rey, 1987; Shibutani et al., 1984; Tehovnik and Lee, 1993; Thier and Andersen, 1998).

Just as in the hypothalamus, in the oculomotor system meaningful behavior was evoked when the stimulation was applied on a behavioral timescale. A saccadic eye movement is so rapid that it is over within 40 to 80 ms depending on the length of the saccade trajectory. Thus a short stimulation train, on the same order of magnitude as the behavior, was used to evoke the complete eye movement (Bruce et al., 1985; Robinson, 1972; Schiller and Stryker, 1972). It was later shown that even shorter trains applied to the superior colliculus, such as 20-ms stimulation trains, evoked truncated saccades (Stanford et al., 1996). Although a saccadic eye movement is normally brief, a gaze shift that includes the eye and the head requires more time to unfold. In experiments in which the monkey's head was free to move, stimulation of oculomotor areas for half a second or longer evoked a coordinated shift of the head and the eyes closely resembling a natural gaze shift (Chen and Walton, 2005; Freedman et al., 1996; Martinez-Trujillo et al., 2003). Likewise, Gottlieb et al. (1993) evoked slow, smooth movements of the eyes, resembling normal smooth pursuit movements, by stimulating in a specific cortical area for half a second or longer.

In the studies cited above, the details of an evoked eye movement depended somewhat on the parameters of stimulation. For example, in the superior colliculus, faster saccades were evoked when the stimulation train had a higher pulse frequency (Stanford et al., 1996). In the FEF, a temporal pattern of pulses that accelerated from a low frequency to a high frequency evoked saccades more effectively, at lower current, than a reversed temporal pattern, a constant frequency, or a randomized temporal pattern (Kimmel and Moore, 2007). The effect of stimulation parameters, however, should not be overstated. Despite these changes in the parameters, the essential movement remains unchanged. No matter what the parameters, stimulation of a site in the FEF or the superior colliculus evokes a saccade in a fixed direction that approximates a natural movement. One cannot, for example, turn a saccade site into a smooth pursuit site, or a horizontal saccade site into a vertical one, or an eye movement site into a foot movement site, by altering the stimulation parameters. Instead, each site has associated with it a specific, coordinated action. Activation of that site, even within a broad range of temporal patterns, causes that action. These findings suggest that the behavioral function of a site is defined largely by its spatial pattern of connectivity, such that activity at that site, in whatever temporal pattern, spreads outward through the same connectivity and produces a relatively consistent behavioral effect.

Seidemann et al. (2002) studied the spread of stimulation-evoked signal in the FEF of the monkey. In this experiment, a stimulating electrode was inserted at one location in the FEF while the entire FEF was optically imaged using a voltage-sensitive dye that glows when nearby neurons become active. The spread of signal from the electrode into the rest of the FEF could therefore be measured. Two surprises emerged from the study. First, even though a specific saccadic eye movement was evoked as expected, the stimulation caused a transsynaptic spread of signal over most of the frontal eye field. The activity did not remain local. Second, the increase in activity was followed after the stimulation train by a drop in activity indicating a profound after-inhibition.

The first surprise finding is easier to understand. Although stimulation of a site evokes a specific eye movement, it is not correct to imagine that the activity at that site alone evokes that eye movement. Activity at that site addresses the entire network, and the movement representation is a highly distributed one. Even with respect to naturally occurring saccades, neurons in the FEF are broadly tuned, and therefore most of the FEF is simultaneously active. It remains perhaps the most subtle and confusing aspect of electrical stimulation, and of brain function in general, that the activity at one location may have a specific function, but not in isolation of the rest of the system; its function is achieved by way of the rest of the system.

The second surprising finding from the Seidemann et al. (2002) experiment, the profound after-inhibition, is more problematical for the electrical stimulation technique. Seidemann et al. pointed out that neurons in the FEF are highly active during a naturally occurring saccade and then become inhibited for a brief time after the saccade. It is possible that the inhibition that follows electrical stimulation is merely an exaggeration of this normal dynamic property of the system. However, the inhibition observed in the optical imaging experiment was so profound and widespread that it sounds a general warning about the technique. Although stimulation may help to reveal function, it should also be taken cautiously because the spread of signal from the stimulated site is unlikely to exactly mimic normal signal spread and probably evokes highly abnormal network dynamics.

STIMULATION TO STUDY PERCEPTION

Electrical stimulation of a sensory brain area causes sensory perception. This simple and ubiquitous truth provides a powerful argument for the technique as a probe of function. Newsome and colleagues (Salzman et al., 1990) adapted the technique to study the extrastriate visual cortex in monkeys. They found that stimulation of targeted sites in cortex could alter the monkey's judgments about visual stimuli. In their experiment, monkeys were trained to watch a moving visual stimulus for one second. At the same time that the visual stimulus was presented, sites in visual cortex were electrically stimulated. When the visual stimulus ended, the monkeys then made a response to indicate the perceived direction of motion of the stimulus. By stimulating specific cortical sites, the experimenters could bias the monkey's judgment. For example, at some cortical sites, the neurons were tuned to upward visual motion. When the experimenters stimulated such a site at the same time that the monkey viewed a downward moving stimulus, the monkey tended to make an error and report that the stimulus was moving upward. The stimulation therefore altered the monkey's perceptual judgment. The results were important because they showed for the first time that neuronal activity in a cortical visual area was not only correlated with perception, but also participated in causing perception.

In a similar experiment, Romo et al. (1998) trained monkeys to perform a tactile discrimination task with the fingertips. On some trials, rather than applying a tactile stimulus to the fingers, the experimenters applied a half-second train of stimulation pulses to the somatosensory cortex in the finger representation. The animal treated the cortical stimulation as though it were perceptually indistinguishable from the externally applied tactile stimulation. Stimulation of cortex at different frequencies was correctly matched by the monkey to different flutter frequencies applied to the fingers. Once again, the result suggests that the cortical neurons influenced the broader perceptual network in a similar manner whether they were naturally active or artificially stimulated.

Stimulation of primary visual cortex (V1) has long been known to produce phosphenes, or perceived spots of light (for review see Tehovnik et al., 2005). This phenomenon suggests that artificial signal injected into V1 percolates through the perceptual machinery in a manner similar to naturally occurring signals. Recently, Tehovnik et al. (2005) investigated the effect of stimulation in V1 of monkeys. The stimulation seemed to mask or interfere with perception of actual visual stimuli, perhaps because it evoked a competing perceptual spot of light. The diameter of the cortical tissue that was directly activated by the current, and the amount of visual space represented by that tissue, predicted the size of the interference effect in the visual field. The stimulation therefore did not cause a general spread of signal that produced indiscriminant interference. Instead, the effect was highly specific to the normal function of the stimulated tissue.

Tolias et al. (2005) electrically stimulated V1 in monkeys and measured the spread of signal using fMRI. In this experiment, the monkeys were anesthetized, and trains of biphasic stimulation pulses were applied to VI through a microelectrode. The experiment used a range currents and durations. Yet even at the highest levels of stimulation, with currents of 1800 microamps and train durations of 4000 ms, the pattern of activation was specific. The stimulation-evoked activity included a directly stimulated region of tissue around the electrode tip, a larger halo of activity around the electrode presumably activated through lateral connectivity, and disparate regions of activity in extrastriate cortical areas to which V1 is known to project. The pattern of activity suggested that, as expected, the stimulation caused signal to spread through the existing connectivity.

One recent puzzling finding is a profound inhibition of some cortical areas obtained during electrical stimulation of V1 but not during visual stimulation of the retina (Logothetis et al., 2006). This finding, like the finding of Seidemann et al. (2002) in the FEF, suggests that although electrical stimulation of a site in the brain causes signal to spread approximately along the normal, physiological pathways, the dynamics of that signal spread are probably not normal. These abnormal dynamics were demonstrated particularly clearly by Douglas and Martin (1991) who found that stimulation of sites in the cat visual cortex evoked an initial neuronal excitation followed by a surge in inhibition. The effect of stimulation on perception or on behavior should therefore be taken with some caution and may provide only a first approximation to the normal function.

TRADITIONAL STUDIES IN MOTOR CORTEX: A LIMITED PERSPECTIVE ON STIMULATION

As reviewed in Chapters 2 and 3, the primary motor cortex was traditionally believed to contain a map of muscles. Yet surface stimulation of the primary motor cortex revealed a messy, overlapping map (e.g., Penfield and Boldrey, 1937; Woolsey et al., 1952). Could it be that the crude, large electrodes and spreading currents of surface stimulation had failed to reveal the details? Microstimulation seemed the ideal technique to reveal a muscle-by-muscle map arranged on the cortex. Asanuma and colleagues (Asanuma, 1975; Asanuma and Rosen, 1972; Asanuma and Sakata, 1967; Asanuma and Ward, 1971) applied the microstimulation technique to the primary motor cortex. Their approach was essentially one of anatomical tract tracing and was therefore different from the technique used in the hypothalamus, in eye movement structures, or in sensory systems. The question of mapping lent itself to smaller currents and shorter stimulation trains.

The goal of the motor cortex experiments was to stimulate as small a locus in cortex as possible and cause the signal to travel down the pathway to the muscles. One technical difficulty was that the direct spread of current around the electrode might activate many adjacent cortical neurons. A second technical difficulty was that the indirect, transsynaptic spread of signal might activate other cortical neurons at a distance. Both of these problems were seen as sources of blur in the mapping technique. Asanuma and colleagues hoped to minimize these problems by stimulating at threshold, using the lowest possible current and shortest possible pulse train necessary to evoke an effect on the muscles.

The spread of signal through connected networks proved to be a difficult problem to overcome. Even a single pulse of current delivered to the cortex evoked an extensive lateral spread of signal among cortical neurons (Jankowska et al., 1975). As might have been expected, the network was apparently built to cause signal to spread transsynaptically. No method of reducing or altering the stimulation parameters could cause the signal to spread only in the descending direction of interest to motor physiologists (down the spinal cord to the muscles) while avoiding other directions. Stimulation appeared to cause signal to ramify through the existing connectivity. In retrospect this result seems obvious. It seems odd to expect that lower currents or shorter durations would cause signal to spread down the pathway of experimental interest while avoiding other pathways. The assumption at the time, however, was presumably that the descending pathway was stronger, or more robust, or more important, and therefore a very low current might activate it and not the less important pathways. This assumption was wrong. Signals spread downward and also laterally regardless of the stimulation parameters, though undoubtedly the overall amount of signal was reduced with lower currents and shorter durations.

Cheney and Fetz (1985) used an even more punctate method of electrical stimulation to study the mapping from motor cortex to muscles. In this method, discussed in more detail in Chapters 3 and 5, a pulse of current was delivered to the motor cortex while muscle activity was measured. Within 5 to 10 ms of the pulse, an effect could be observed at the muscle. By focusing the experiment on this short latency signal, Cheney and Fetz were able to study the most direct descending pathway regardless of how the signal might percolate through the larger network.

The experiments on motor cortex therefore pioneered the use of electrical stimulation for anatomical tract tracing. In this approach to electrical stimulation, the optimal method is to stimulate one location, record in another location, and measure the latency of the evoked activity. A latency measure at the millisecond time scale is necessary because it indicates whether the signal traveled over a fiber, a monosynaptic pathway, or a polysynaptic pathway. For example, Sommer and Wurtz (2002) used the technique brilliantly to probe connectivity within the oculomotor system. In its original use in the motor system, however, one major difficulty faced the anatomical tract-tracing technique. There is no single pathway from cortex to muscles. The intervening circuitry is a network with many possible paths through which signal can travel. Although the work of Cheney and Fetz (1985) was able to probe the shortest latency route through this network, that route presumably does not represent the full complexity of the network.

In summary, the studies of motor cortex used a philosophy of electrical stimulation markedly different from the philosophy used in the study of other brain systems. In motor cortex, low currents and brief pulse trains were used to perform anatomical tract tracing. Behavioral function was not probed with electrical stimulation, except in the sense that the behavioral function of motor cortex was assumed to be defined by its muscle map. The motor cortex literature was somewhat obsessed with the problem of the spread of signal through connected networks, something that researchers tried to minimize and failed to eliminate, rendering the technique of limited value to them. In other brain systems, in contrast, stimulation was used for a fundamentally different purpose. Parameters were optimized to evoke meaningful behavior. Stimulation trains were typically longer and of higher current. No assumptions were made about the specific pathways through which the signal spread, whether one main pathway or a diffuse network. The transsynaptic spread of signal was not viewed as an inconvenient artifact, but instead viewed as the key process that allowed the experiment to succeed.

6. What Can Be Learned from Electrical Stimulation? 93

STIMULATION CONTROVERSY

Now that some background on electrical stimulation has been described, it should be clear to the reader why our electrical stimulation studies in motor cortex caused umbrage. Our method was borrowed from the oculomotor literature. Dr. Moore, who worked down the hall from me, was electrically stimulating the FEF of monkeys at the beginning of a brilliant set of experiments that demonstrated a causal link between eye movement control and visual attention (Moore and Armstrong, 2003). I was working at the time on sensory signals that could be measured in the motor cortex of monkeys. When Moore, Taylor, and I transplanted the stimulation technique into the motor cortex, we used methods that were a matter of course in the oculomotor literature and yet were frankly bizarre within the motor cortex literature.

We used biphasic pulses rather than negative pulses. The balanced charge allows for high currents and long durations while minimizing electrolytic damage to the brain. We explored a range of frequencies, train durations, and current amplitudes and settled on the values that optimized the evoked behavior. For most stimulation sites, a stimulation train that was half a second long seemed to optimize the movement, allowing the action to completely unfold. Shorter stimulation trains resulted in truncated movements and longer stimulation trains seemed to hold the joints at the terminus of the movement. Half a second approximates the duration of a monkey's normal reaching and grasping. A behaviorally relevant time scale, therefore, seemed to optimize the evoked behavior. We used current amplitudes that ranged from about 20 to 200 microamps. We used pulse frequencies that ranged between 50 and 500 Hz. In general, the evoked movement was similar at different frequencies, with a tendency for higher frequencies to evoke faster movements. Most of our experiments were performed with a 200 Hz pulse frequency, roughly in the middle of the effective range.

To investigators aware of the literature on the hypothalamus, the oculomotor system, and sensory systems, our methods seemed normal. The parameters were more or less midrange. The approach of optimizing the evoked behavior and making no assumptions about the spread of signal through networks was familiar. To motor physiologists, we seemed out of our minds. The currents were five times higher than normally used, the durations ten times longer. Our results could not possibly address the traditional question of motor cortex stimulation, specifically the direct wiring of sites in cortex to muscles. We had no doubt caused a spread of signal through the entire system, and therefore could learn nothing about the specific pathway from the electrode site, down through the spinal cord, to the muscles.

Five concerns were frequently raised by audience members at talks or by reviewers of our papers and therefore I will raise them here as well.

Artificiality of Long Stimulation

A common suggestion was that stimulation for 500 ms was unphysiological; only brief stimulation trains, such as 50 ms trains, could be trusted. Yet one must ask whether a 50-ms stimulation train is any more physiological than a 500-ms train. They are both artificial. During normal movement, such as during a reach or a grasp, neurons in motor cortex do not fire in brief 50-ms bursts. To a first approximation, they are active throughout the movement, which typically lasts on the order of half a second. One could make an argument therefore that stimulation for 500 ms captures the first-order structure of a normal spike train, whereas stimulation for 50 ms captures no aspect of a normal spike train. Probably an even more physiologically relevant approach would use a stimulation train with a complex temporal pattern of pulses taken from the activity of motor cortex neurons themselves. Such an experiment has not yet been done.

To me, however, it is important not to judge a technique by whether it is physiologically natural. The central question is whether insight can be gained from the data, not whether the technique is physiological. For example, damaging the brain obviously causes an extreme disruption of natural physiological function, and yet the lesion technique has resulted in some of the most fundamental insights into brain function.

Damage Caused by Stimulation

A second common criticism was that the long stimulation trains in our experiment must have damaged the cortical tissue around the electrode tip, thereby somehow artificially causing the unexpected results in motor cortex. When a train of unipolar, negative pulses is applied to the cortex, the neurons around the electrode tip can be killed (Asanuma and Arnold, 1975). This damage can be detected in several ways. The stimulation-evoked movement will cease to occur after a few stimulation trials because the tissue around the electrode tip has died. Neuronal signals will no longer be detectable through the electrode at that placement. Histological processing will reveal a small damaged sphere of tissue. Our experiments, however, used biphasic pulses, balancing the negative and positive charge. As a result little or no damage was expected. Stimulation at a site evoked a consistent movement even after hundreds of trials. After testing the effect of stimulation, neuronal signals could always be obtained at that site through the electrode. Histological processing revealed no damaged tissue except the expected tracks of gliosis caused by the electrode penetration itself.

Connections Versus Function

A third common criticism was that the stimulation technique is traditionally used for tracing anatomical connections and totally unsuited to studying function. In that view, our attempt to use the method to study function was inappropriate. Yet given the literature on stimulation reviewed above, the technique appears to be particularly effective at probing behavioral function wherever it is used in the brain. Although it can be used to trace anatomical pathways, this use is subject to more interpretational difficulties because most

6. What Can Be Learned from Electrical Stimulation?

connectivity in the central nervous system resembles a network rather than a pathway.

Activating Widespread Networks

A common concern was that the stimulation must have affected many connected structures. How does one know if an evoked movement is truly a function of the directly stimulated neurons around the electrode tip, or instead a function of the structures connected to those neurons? Perhaps the evoked movement is actually caused by spinal circuits, basal ganglia circuits, cerebellar circuits, or other motor circuits.

In a similar view, Strick and colleagues (Slovin et al., 2003) pointed out that stimulation of one site in motor cortex causes neuronal signals to spread transsynaptically to other locations in cortex, presumably via the rich lateral connectivity within the motor cortex. Given this spread of signal, how can one infer anything about the functions of the local neurons at the electrode tip?

In the use of electrical stimulation, it is necessary to distinguish two kinds of signal spread. Direct spread, sometimes called passive spread, is the spread of the electrical field around the electrode tip that activates neighboring neurons. Indirect or active spread is the spread of neuronal signals across synapses and through networks. Ideally the passive spread is minimized, or at least restricted to the experimentally targeted neurons. The active spread, the percolation of signal through connected networks, is the goal of the technique, allowing function to be probed. It is not an error or artifact to be avoided. The evoked movement is a function of the directly stimulated neurons because of their effect on connected structures. The most basic truth of the brain is that no neuron has a function by itself. Its function is defined by its connections with, and therefore its influence on, other neurons.

Single-Neuron Recording Versus Stimulation

Perhaps the most familiar criticism of our stimulation method lay in its comparison to single-neuron recording. According to the criticism, electrical stimulation is suspect because the signal, injected at one point in the cortex, spreads in some unknown fashion through the circuitry. In contrast, single-neuron recording is a controlled, local experiment that avoids the spread of signal through networks. Yet this suggestion is not correct. Single-neuron recording is not local. When one measures action potentials in a neuron, although the measured signal is initially local, it spreads transsynaptically, affecting other neurons, causing a ramification of signal in some unknown fashion through the circuitry. Whether one measures a local signal or injects a local signal, it does not remain local but spreads through the existing circuitry. This non-locality is not a property of one or another technique. It is the single most fundamental property of the brain.

Chapter 7

Complex Movements Evoked by Electrical Stimulation of Motor Cortex

INTRODUCTION

This chapter summarizes our electrical stimulation studies in the macaque motor cortex, in which stimulation on a behaviorally relevant time scale evoked complex movements that resembled behaviorally meaningful actions. The chapter begins with a brief summary of methods. A technical description of our methods can be found in our experimental papers (e.g., Graziano et al., 2005; Graziano, Taylor, et al., 2002). Here the purpose is to provide a useful description for those not directly familiar with electrical stimulation in the monkey brain. The chapter then summarizes two sets of findings, first the quantitative profile of the evoked arm and hand movements, and second the manner in which evoked movements resemble actions in a monkey's natural behavioral repertoire. The chapter ends with a review of other experiments using similar stimulation techniques to study motor areas in a range of animal species.

METHODS

The methods that we used are standard for all institutions throughout the world that study brain function in monkeys. The experiments were done under the supervision of an attending veterinarian, and were in accordance with federal regulations for animal experimentation. Typically at the end of an experiment all surgically implanted probes were removed and the animals were sent to a wildlife sanctuary. The studies summarized in this chapter included seven monkeys, in all of whom a similar pattern was obtained. The monkeys were first trained to climb out of the home cage and into a primate chair in which they were restrained by means of a collar for two or three hours per day. The chair was open at the front and sides, allowing for almost total range of movement of the arms. In some experiments, one arm was strapped down to allow us to focus the experiment on the other arm. The animal was not trained on any specific task. It was acclimated to the chair and to the experimenters through daily training sessions over a period of weeks. When the animal was calm enough to reach toward and eat small fruit rewards, it was considered sufficiently trained to begin the experiment.

In a sterile surgery under anesthesia, the animal's head was implanted with a cap made of dental acrylic. Two metal items were embedded in the acrylic. First, a metal bolt extended from the back of the implant, providing an anchor for holding the head still. Second, a metal cylinder with a screw top was embedded in the acrylic over the location of motor cortex. Inside the metal cylinder, a small hole (approximately 1 cm in diameter) was drilled through the skull, exposing the dura, the membrane that encases the brain. The animal was allowed three weeks to fully recover from this surgery.

During the daily experiment sessions, the monkey sat in a primate chair with its head restrained by the head bolt. The screw top on the metal cylinder was removed and physiological saline was poured into the cylinder to maintain the moistness of the dura. A hydraulic microdrive was then mounted to the cylinder. By means of this microdrive, a syringe needle was lowered until its point touched the dura. The syringe needle served the purpose of protecting the electrode, which was housed inside the needle. Once the needle was in place over the dura, the electrode was advanced such that its tip protruded out of the syringe needle, through the dura and into the brain. Because the brain contains no sensory receptors, this introduction of an electrode into the brain is undetectable by the monkey.

Our electrodes were standard, high-impedance electrodes for single-neuron recording. They were made of varnish-coated tungsten with approximately 20 microns of the tungsten tip exposed. The electrical impedance of the tip usually began high, approximately 3 to 5 MOHM, but through repeated use of the electrode, inserting it through the dura at different locations, the varnish would begin to wear at the tip and the impedance would drop sometimes as low as 0.5 MOHM. When the impedance was high, the electrode was particularly good for recording the activity of individual neurons, but was not ideal for electrical stimulation because of the difficulty of reliably passing a current through the high-impedance device. When the impedance was low, the electrode was poor at monitoring neuronal activity but was ideal for electrical stimulation. Therefore depending on the focus of the experiment for that day, we tended to use electrodes of relatively higher or lower impedance.

While lowering the electrode into the brain, we monitored the amplified neuronal activity over a loud speaker and on an oscilloscope. In this manner, the depth at which the brain was first reached could be determined by the appearance of cellular activity, and the depth at which the electrode penetrated beyond the cortex and entered the underlying white matter could be determined by the drop in neuronal activity.

When the electrode was placed within the motor cortex, we switched the wires from a recording configuration to a stimulating configuration. The electrode served as the positive pole, and the guard tube for the electrode, placed over the dura and immersed in saline, served as the negative pole. After testing the effects of stimulation at a site, we then switched back to the recording configuration to confirm that neuronal signals could still be detected, indicating that the stimulation had not killed the neurons. We then lowered the electrode to the next site for study. We typically studied one to three depths within the

7. Complex Movements Evoked by Electrical Stimulation

cortex, each separated by 0.5 mm. If the electrode was advanced perpendicularly to the cortical surface, the effects of electrical stimulation were similar at different depths. Movements could be evoked with smaller currents in the deeper layers of cortex, consistent with the known properties of motor cortex, in which the deeper layers contain large output neurons.

The current entering the brain was directly monitored at all times. This direct monitoring was necessary because, regardless of the intended stimulation parameters, the impedance of the electrode could sometimes alter the current flow. We adjusted our stimulator until the desired current was confirmed. It was particularly important to confirm that the negative and positive phases on the biphasic pulse were balanced, or else a buildup of charge might damage the brain around the electrode tip.

Pulse Width

Figure 7-1 shows the stimulation parameters typically used in our experiments. This diagram shows a train of biphasic pulses with the negative phase leading,

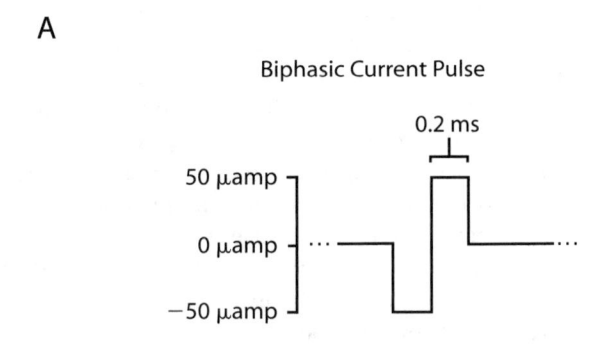

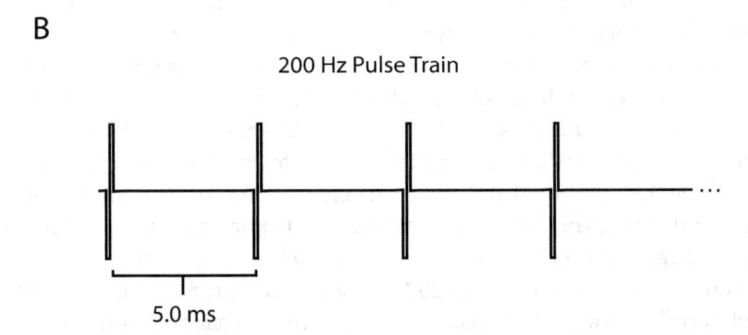

Figure 7-1 Stimulation parameters used in our experiments on monkey motor cortex. **A.** Biphasic, negative-leading pulse with 0.2-ms pulse width and 50-microamp current. **B.** Train of pulses presented at 200 Hz. Stimulation trains were typically 500 ms in duration (100 pulses).

each phase 0.2 ms in duration. Wider pulses generally result in more effective stimulation (e.g., Ranck, 1974; Tehovnik, 1996; Tehovnik et al., 2006). For cortical stimulation, the effectiveness asymptotes at about 0.4 ms pulse width. Most of our initial studies used 0.2-ms pulse widths but in our later experiments we more often used 0.4-ms pulse widths.

Pulse Frequency

The frequency of pulses shown in Figure 7-1 is 200 Hz, corresponding to an interpulse interval of 5 ms. We found that a similar movement could be evoked with a range of frequencies between about 50 and 500 Hz. Outside of this range, the evoked movement was weak or absent. Within this range, although the movement was similar at different frequencies, the speed of the movement tended to increase with increasing frequency, an effect that we quantified for the case of stimulation-evoked hand movement (Graziano et al., 2005). We used a frequency of 200 Hz for most experiments because it was near the center of the effective range.

Pulse Amplitude

The amplitude of pulses shown in Figure 7-1 is 50 microamps, a current commonly used in our experiments. At each site we first obtained the threshold current at which a movement was visible on 50% of trials. These thresholds, as discussed in Chapter 3, suffer from the tip-of-the-iceberg phenomenon. Any complex movement recruits many muscles to differing extents. The definition of *threshold current* is the current at which only the one or few most strongly activated muscles remain detectably above the noise. As a result, the movement evoked at threshold is always a simple twitch, typically of one or two joints. We obtained the lowest thresholds, as low as 3 microamps, in the primary motor finger representation, as expected. Thresholds throughout the rest of motor cortex ranged from about 10 to 30 microamps.

Once the threshold was determined for a stimulation site, we then increased the current until a coherent action was evoked. Because there is no clear definition of a coherent action, there was no hard current threshold. Typically, complex movements were clearly observable at currents between about 20 and 100 microamps. Some types of actions, such as the defensive movements, could be evoked robustly at the lower end of this range, whereas other actions, such as the climbing-like movements, tended to require more current. Once a complex action was identified for a cortical site, the components of the movement could often be detected at low current. For example, a 50-microamp current might reveal a hand-to-mouth site. Once this action was observed, we could then lower the current to 20 microamps and observe a movement of the hand and the mouth that was consistent with a weakened form of the hand-to-mouth action.

It has been suggested that our currents were atypically high (Strick, 2002). This suggestion appears to be based mainly on the motor cortex literature.

7. Complex Movements Evoked by Electrical Stimulation

Our currents are in the range typically used in studies of cortical stimulation. Others have measured the spread of current through cortex (Ranck, 1974; Tehovnik, 1996; Tehovnik et al., 2006), and based on the response curves obtained in these studies, the area of directly activated cortex in our experiments is estimated to be within a 0.25-mm radius.

Train Duration

We tested a range of durations for the pulse train, from 20 ms to 2000 ms. The train duration was the single most important factor in obtaining a coherent movement. Short durations evoked muscle twitches of no obvious behavioral significance. Longer stimulation allowed the movement to unfold. Most actions were recognizably complete at a duration of 500 ms. Even longer durations sometimes resulted in the relevant joints freezing at the final posture. After extensive exploration of different durations in our initial experiments, we settled on trains of 500-ms duration. One monkey had unusually fast natural movements, and stimulation in this monkey appeared to evoke complete movements in approximately 400 ms. Therefore in experiments on this particular monkey we typically used a 400-ms train duration.

Optimizing for Evoked Behavior

The parameters discussed above were chosen because empirically they resulted in coherent movements that resembled parts of the monkey's natural repertoire. The parameters were not optimized to limit the spread of neuronal signal, focus the signal onto the cortico-spinal pathway, or produce any other hypothetical effect. The goal was to optimize the observable movement and then to describe those evoked movements.

RESULTS 1: QUANTITATIVE DESCRIPTION OF EVOKED MOVEMENTS

Convergence

A property obtained at almost all stimulation sites whether in primary motor cortex, premotor cortex, or SMA was a convergence of the joints to a specific final configuration. This convergence was most noticeable with respect to the arm. Stimulation in the forelimb representation in the lateral motor cortex almost always drove the joints of the shoulder, elbow, and wrist at least partially toward some final set of angles regardless of the starting configuration. As a result, the hand almost always moved toward a final region of space regardless of its starting location. Stimulation at a hand-to-mouth site, for example, caused the arm to move to a specific configuration, resulting in the hand arriving at and staying at the mouth. We also saw convergence in other body parts, but this convergence was more difficult to quantify. Stimulation in the mouth representation, for example, sometimes drove the jaw, lips, and tongue into a specific configuration regardless of the starting configuration.

Figure 7-2 shows the spatial convergence of the hand for fourteen example sites in the lateral motor cortex. We used a camera system (Optotrack 3020) to track the position of an infrared light-emitting diode fixed to the hand. Figure 7-2b shows an example in which the hand converged toward the mouth during the half second of electrical stimulation. Each line in the figure shows the

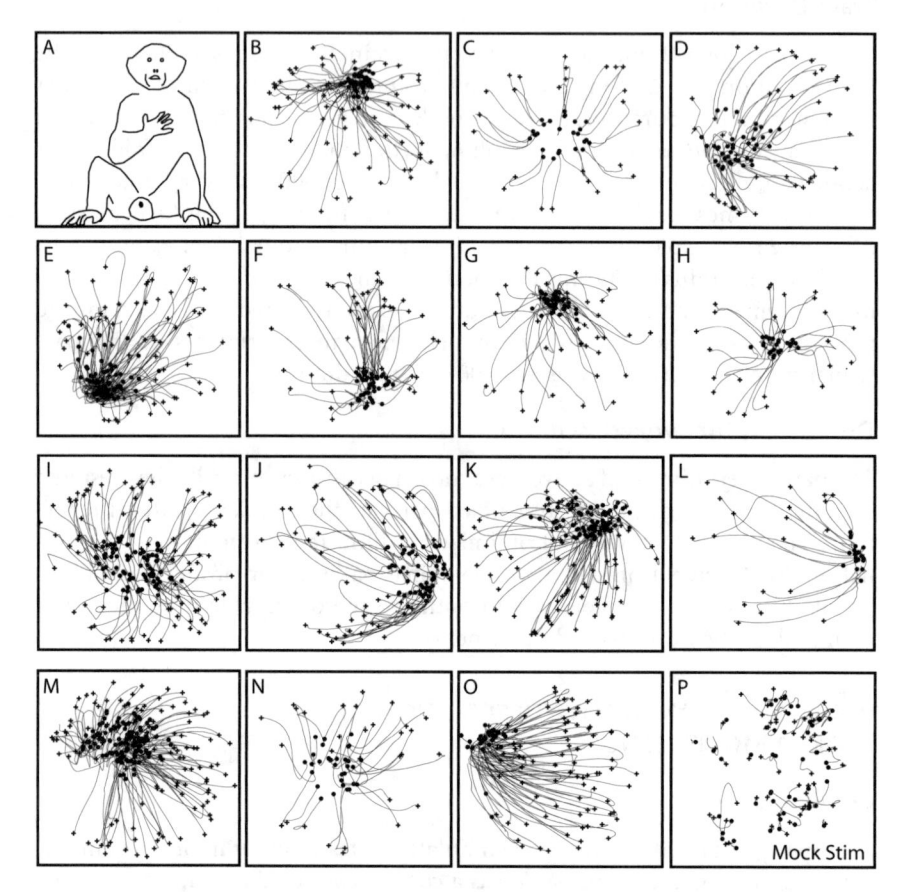

Figure 7-2 Examples of hand movements evoked by microstimulation in motor cortex. **A.** The monkey drawing indicates the approximate size, location, and perspective of the monkey within the square frame. The height of the frame represents 50 cm. **B–O.** Stimulation-evoked hand movements from 14 typical stimulation sites. Each thin black line shows the path of the hand during a stimulation train. The + indicates the start of the movement. The black dot indicates the end of the movement. In a small number of trials, the tracking markers were transiently blocked from the view of the camera due to the specific posture of the limb. In these cases, the trace is interrupted. **P.** Result of mock stimulation in which the wires to the electrode were disconnected but all other aspects of the testing were the same. The traces in panel O show the result for this same cortical site when the wires were connected. Adapted from Graziano et al. (2005).

7. Complex Movements Evoked by Electrical Stimulation 103

path of the hand from the onset of stimulation (indicated by a +) to the offset (indicated by a dot). For 100% of the ninety-one tested stimulation sites in motor cortex, the distribution of hand positions in space was significantly smaller at the end of the stimulation train than at the beginning. In contrast, Figure 7-2p shows the effect of mock stimulation, in which the protocol was identical but the wires connecting the stimulator to the brain were disconnected. In this case, no significant convergence was obtained, indicating that the movement was not a voluntary movement shaped by contextual training.

Some stimulation sites showed a more robust convergence than others. For example, hand-to-mouth sites almost always involved a movement of the hand to a tight cluster within a few centimeters of the mouth. Movements of the hand to central space just in front of the chest, in contrast, often involved a large, distributed terminal field of hand positions. These differences can be seen among the examples in Figure 7-2. Even in those cases when stimulation evoked a messy movement that, to the experimenter's eye, did not appear to converge, on analysis the convergence of the hand to a tighter spatial distribution was statistically significant.

Map of Hand Positions Arranged Across the Cortical Surface

As described in the previous section, stimulation of each cortical site tended to drive the arm to a specific final posture or joint angle configuration. As a result, the hand tended to move to a final location in space. These evoked hand locations were not randomly intermingled across the cortical surface. Instead they formed a rough map across the arm and hand representation of the lateral motor cortex. Ventral sites in cortex were associated with upper hand locations, dorsal sites with lower hand locations, anterior sites with lateral hand locations, and posterior sites with medial hand locations. This mapping is shown for one monkey in Figure 7-3. The mapping is noisy and of a statistical nature. Like the somatotopic map (see Chapter 3), the map of hand location contains considerable overlap and intermingling. Yet a broad organization is nonetheless apparent. The vertical height of the hand was most consistently mapped among monkeys. In some monkeys, an anterior and dorsal area was tested that probably encroached onto SMA. In this area, the statistical map broke down, and upper and lower hand positions were intermingled.

Smooth Speed Profile of the Hand

Normal movements of the hand follow a typical bell-shaped speed profile, in which the peak speed is roughly linearly correlated with the length of the hand movement (e.g., Bizzi and Mussa-Ivaldi, 1989; Flash and Hogan, 1985). An important initial question therefore was whether the stimulation-evoked movements had these basic properties of normal movement or whether they were irregular and jittery spasms. Figure 7-4A shows speed profiles for a typical stimulation site, aligned on stimulation onset. These profiles are roughly bell shaped. This bell-shaped property can be seen more clearly in Figure 7-4B

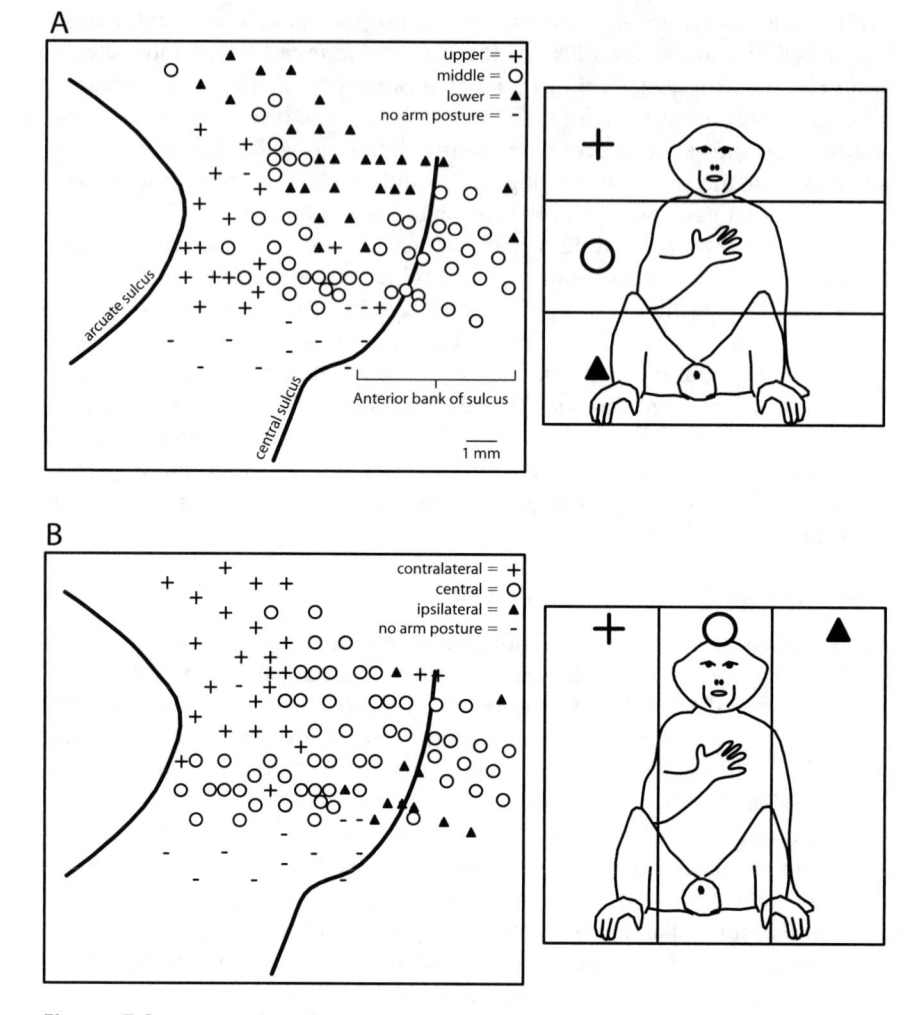

Figure 7-3 Topography of stimulation-evoked hand locations in the motor cortex. Sites plotted to the right of the central sulcus were located in the anterior bank of the sulcus. **A.** Distribution of hand positions along the vertical axis, in upper, middle, and lower space. Each site was categorized based on the center of the range of evoked final positions. Height categories were defined as follows: lower = 0 to 12 cm from bottom of monkey, middle = 12 to 24 cm, upper = 24 to 36 cm. Dashes show electrode penetrations where no arm postures were found; usually the postures from these locations involved the mouth or face. **B.** Distribution of hand positions along the horizontal axis, in the space contralateral, central, or ipsilateral to the stimulating electrode. Horizontal categories were defined as follows: contralateral = 6 to 18 cm contralateral to midline, central = within 6 cm of midline (central 12 cm of space), ipsilateral = 6 to 18 cm ipsilateral to midline. Adapted from Graziano, Taylor, et al. (2002).

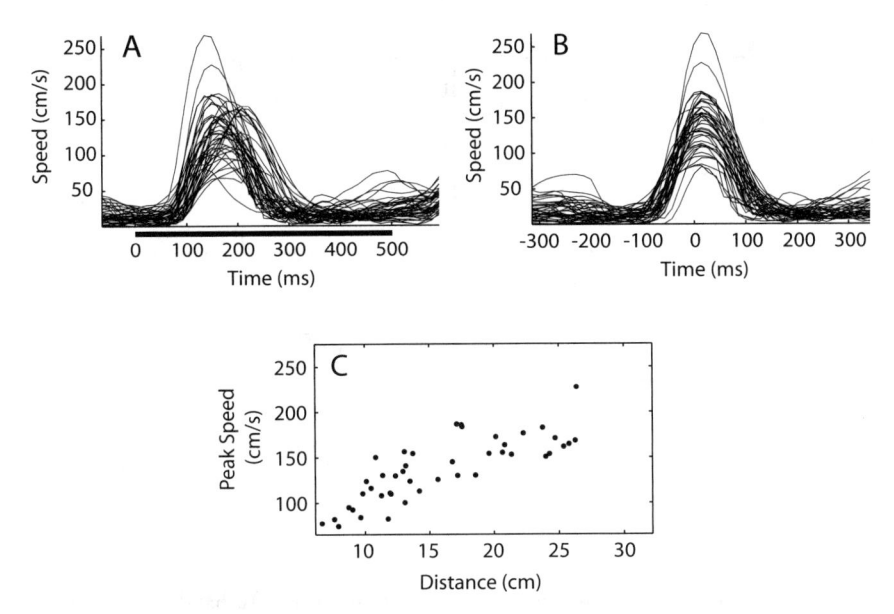

7. Complex Movements Evoked by Electrical Stimulation

Figure 7-4 Speed profiles for a typical stimulation site. **A.** Hand speed as a function of time during stimulation. Each trace shows the result for one stimulation trial. Speed measured in 14.3-ms increments. Thick black bar at bottom shows the time of the stimulation. **B.** Same data as in A, but the traces are aligned on the time of peak speed. **C.** Peak speed during the stimulation-evoked movement as a function of the distance that the hand traveled. Adapted from Graziano et al. (2005).

in which the same trials are aligned on peak speed. The mean speed profile for this set of trials significantly matches a Gaussian curve (regression analysis, $F=405$, $p<0.0001$). Figure 7-4C shows the peak speed as a function of the distance that the hand traveled. The relationship is roughly linear, in which greater peak speeds occurred during longer movements. For all ninety-one sites tested in this fashion, the speed profile significantly fit a Gaussian curve, and for ninety of the ninety-one sites a significant linear trend was obtained between peak speed and movement distance. The evoked movements therefore were not jagged or spasmodic movements as might occur in an epileptic fit but instead resembled normal movements in their speed profile.

We typically tested sites using 200 Hz stimulation, a frequency borrowed from the oculomotor literature (e.g., Robinson and Fuchs, 1969). For some cortical sites, we measured the effect of different frequencies. Figure 7-5 shows the results for one site tested with stimulation at 100 Hz, 150 Hz, 200 Hz, and 250 Hz. The movement of the hand was similar across these different stimulation frequencies. In each case, the hand converged from a range of initial positions toward a similar final region of space. As shown in Figure 7-5E, the peak speed of the movement varied somewhat with stimulation frequency. The lowest speeds were obtained with 100-Hz stimulation; the highest speeds were

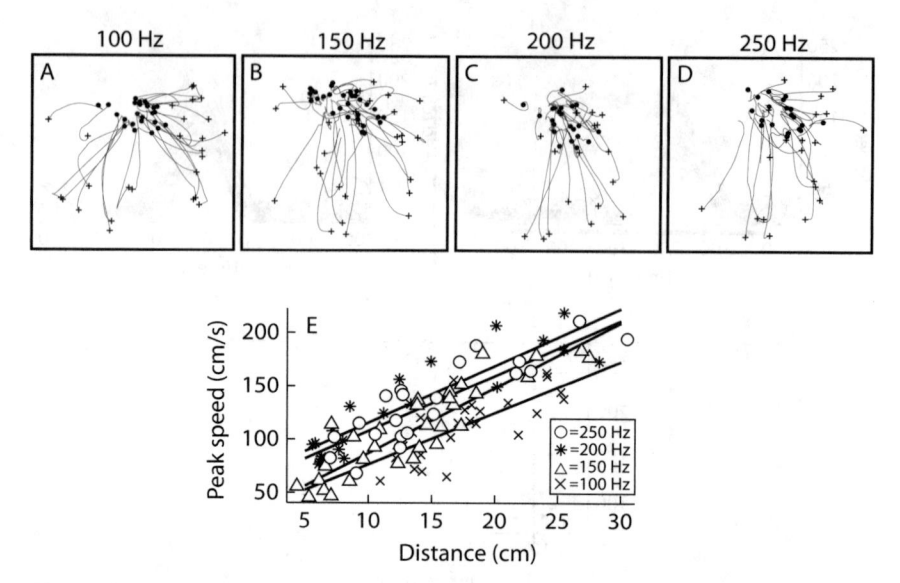

Figure 7-5 Effect of different stimulation frequencies on evoked hand movement for one stimulation site. **A–D**. Movement of the hand evoked by stimulation at 100, 150, 200, and 250 Hz. **E**. Peak speed of movement versus distance that the hand moved for the four different stimulation frequencies. Adapted from Graziano et al. (2005).

obtained with 200-Hz and 250-Hz stimulation. This relationship between stimulation frequency and hand speed was significant (multiple linear regression, $F=15.53$, $p<0.0001$). These results suggest that similar movements can be evoked with a broad range of stimulation frequencies, and that higher frequencies tended to evoke somewhat higher speeds.

Interaction Between Joints Stabilizes the Hand for Some Stimulation Sites

One hypothesis about the evoked movements is that stimulation independently drives each joint to a specific angle, and as a result of this aggregate of joint angles, the hand moves to a location in space. Another hypothesis is that the joints move in a coordinated manner, each joint adjusting to compensate for slight deviations in the other joints, to bring the hand more specifically toward a desired location. We asked whether the trial-by-trial variability in each joint might compensate for the variability in the other joints in a manner that would help to stabilize the hand in space. To test this hypothesis we tracked the location of multiple points on the arm and reconstructed four arm angles: shoulder elevation, shoulder azimuth, shoulder internal/external rotation, and elbow flexion. These arm angles, together with the lengths of the arm segments, define the position of the wrist in space.

Figure 7-6 shows the results for one example site. For each of the thirty-seven trials, using the angles reached at the end of stimulation and applying

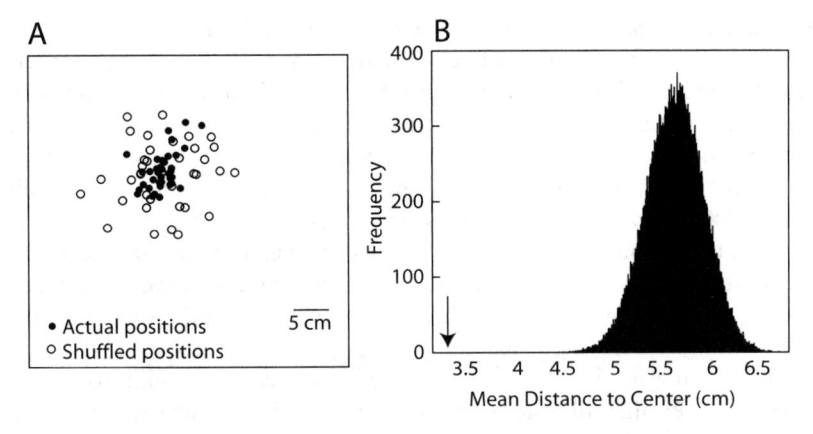

7. Complex Movements Evoked by Electrical Stimulation

Figure 7-6 Interactions between joints stabilized the hand position. **A.** Data from one example site. Black dots show final hand positions for 37 stimulation trials, calculated from the measured joint angles. Open circles show final hand positions calculated from 37 "shuffled" trials, using the same data but with the joint angles randomly shuffled across trials. **B.** A total of 40,000 different shuffles of the 37 trials was tested. For each shuffle, the 37 final hand positions were found through forward kinematics and the mean-distance-to-center was calculated (a measure of the spatial spread of hand positions). These mean-distance-to-centers were then plotted on the frequency histogram shown. The mean-distance-to-center for the actual data is indicated by the arrow. Adapted from Graziano et al. (2005).

forward kinematics, we calculated the final position of the wrist. These final wrist positions are plotted as black dots in Figure 7-6A.

We then randomly "shuffled" the trials in the following manner. A shuffled trial might contain the shoulder elevation angle from Trial 1, the elbow flexion angle from Trial 16, etc. The rule for a shuffled trial was that none of the four joint angles that composed the shuffled trial had been collected on the same actual trial. In this fashion, thirty-seven randomly shuffled trials were constructed. We used forward kinematics on these shuffled trials to calculate final wrist positions. If the joints seek their final angles independently, then shuffling the trials in this manner should have little effect on the result. However, if the joints normally interact within a trial, such that a slight deviation in one joint is compensated by slight deviations in the other joints, then shuffling the trials will remove this interjoint compensation and result in a wider distribution of final wrist positions. As shown in Figure 7-6A, the shuffled trials (open circles) did show a wider distribution of final wrist positions than the actual trials (filled circles). As a measure of the spread of final positions, we used the mean distance to the centroid of the cluster. For this example site, the mean distance to the centroid for the shuffled trials was 6.03 cm and the mean distance to the centroid for actual trials was 3.33 cm.

Figure 7-6A shows only one possible reshuffling of the thirty-seven trials. We tried forty thousand possible reshuffles. Of these many ways to reshuffle the data, all of them resulted in a greater spread of final hand positions than the actual data ($p < 0.0001$ based on Z score). Figure 7-6B shows the distribution for the forty thousand shuffles and for the actual data. These results show that within a stimulation trial, the joint angles were not independent but instead interacted. Deviations in some joints must have been matched by compensatory deviations in other joints, in a manner that helped to stabilize the hand at a particular location in space.

Of sixty-one sites tested in this manner, 54% had a significant effect similar to that shown in Figure 7-6. For 46% of the sites, no significant effect of joint interaction was obtained. Thus only approximately half of the sites showed this stabilization of the hand in space.

This finding of electrically evoked hand stabilization should not be over-interpreted. Presumably the motor cortex controls a great variety of movement parameters. The results suggest that hand location is at least one relevant control variable for the electrically activated circuits. The value of the finding is that it shows that electrical stimulation does not simply contract muscles or rotate joints in a fixed manner but does so with some degree of sophistication that may be behaviorally relevant.

Some Stimulation Sites Compensate for a Weight on the Hand

For almost all cortical sites, stimulation caused the hand to converge from any initial position toward a final region of space. How is this final position affected by a weight fixed to the arm? We tested sites by weighting the wrist with a 90-g lead bracelet, comprising approximately 25% of the arm's weight.

Figure 7-7 shows data from one example site. Figure 7-7A shows trials in which no weight was fixed to the arm, and Figure 7-7B shows interleaved blocks of trials in which the hand was weighted with the 90-g lead bracelet. In both conditions, the hand stabilized at a similar final height by the end of stimulation. The weight appeared to affect the hand initially, pulling it to a slightly lower position, but in the second half of the trial the hand rose up to a similar final height as without the weight.

One potential concern with the above test is that the weight may be too small to have a reliable effect on the height of the hand, whether or not compensation is present. We used a mathematical model to approximate the expected effect of a 90-g weight on the movement of the arm assuming no compensation for the weight, and assuming spring-like properties for the muscles. Figure 7-7C shows the simulated trajectories obtained from the model, in the condition that no weight is present. The model does a good job of simulating the actual data shown in Figure 7-7A, because the spring constants in the model were adjusted to optimize the fit with the unweighted data. Figure 7-7D shows the simulated trajectories obtained from the same model in the condition that the weight is added to the wrist. In the model, the weighted hand converged on a lower position, on average 5.8 cm lower than in the actual data. This result suggests that without compensation, the weight is expected to pull the hand to a lower position. Yet the electrically evoked

7. Complex Movements Evoked by Electrical Stimulation

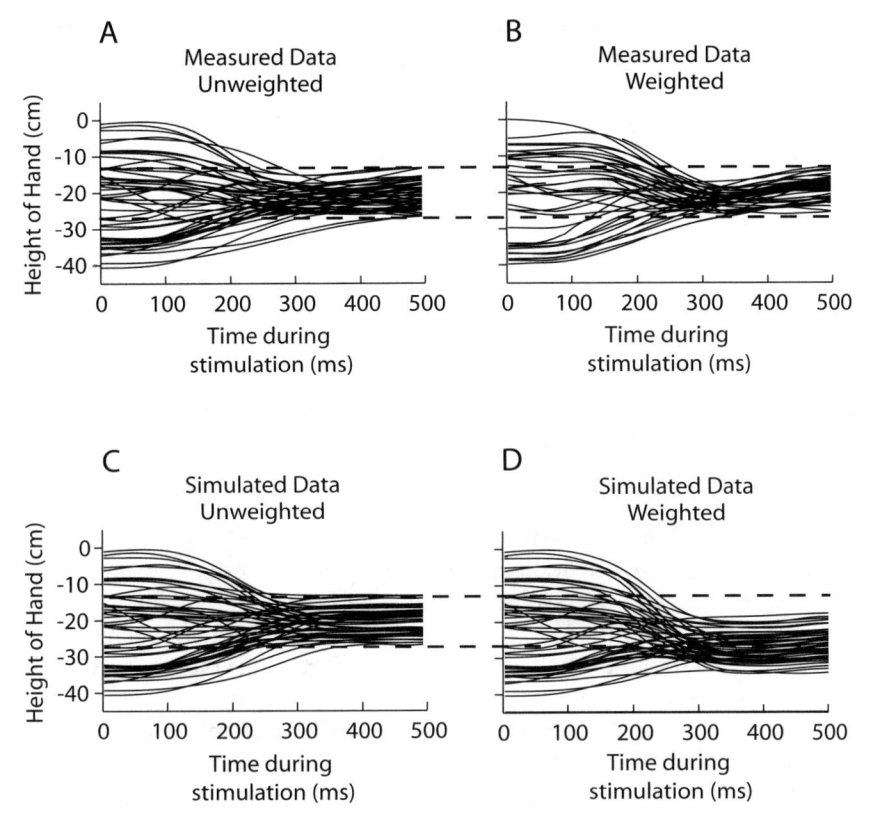

Figure 7-7 Effect of a weighted bracelet on the height of the hand. **A.** Data from an example site. Y axis shows height of hand relative to mouth height, X axis shows time during stimulation trial. Each trace shows data from one trial. Dotted lines show the range (min and max) of final heights. **B.** Data from the same example site as in A. On these trials, a 90 g weight was fixed to the wrist. **C.** We modeled the physics of the arm and used the data from the trials shown in A to find a best fit for the spring constants and damping forces in the model. The graph shows the calculated trajectories for the model. **D.** The calculated trajectories based on the model, when a 90-g weight was added to the wrist in the model. Adapted from Graziano et al. (2005).

movements converged on the same position. Thus the electrically stimulated circuitry was apparently able to compensate for the weight, ultimately lifting the hand to approximately the same final height as without the weight. Of the fifty sites tested, forty-seven (94%) showed a similar compensation for the weight.

Patterns of Muscle Activity During Stimulation-Evoked Movements

In the previous sections, the stimulation-evoked movements were described as sharing many features of the monkey's normal behavioral repertoire. A particularly important question is whether the patterns of muscle activity evoked by stimulation are similar to the patterns of muscle activation during normal movements. The short answer is that they are similar but not the same. The evolution of muscle activity over time is not normal during the extended stimulation train. Much like the study by Seidemann et al. (2002) in the monkey FEF and the study by Douglas and Martin (1991) in the cat visual cortex, reviewed in Chapter 6, the stimulation trains in the motor cortex appear to mimic some aspects of normal function but produce abnormal temporal dynamics.

We measured the electromyographic (EMG) activity of several limb muscles including the biceps and triceps (that help actuate the elbow), and the deltoid and pectoralis (that help actuate the shoulder) during stimulation of motor cortex.

In one study we fixed the limb in a holder and stimulated the motor cortex (Graziano, Patel, et al., 2004). The EMG activity for each muscle typically rose rapidly at the onset of stimulation, remained approximately at a plateau level throughout the duration of the stimulation train, and then dropped back to baseline activity after the offset of the stimulation train (Figure 7-8A). This type of tonic muscle activity is not typical of normal movement.

Arguably, by fixing the limb in a holder, we placed the limb in an isometric condition, not a movement condition. During normal isometric application of force, muscle activity does indeed rise to a plateau and remain at the plateau during the application of the force. In the next experiment, therefore, we released the arm from the holder and recorded EMG while the arm was free to move (Taylor et al., 2002). Under this condition, during stimulation of motor cortex, the muscle EMG was less obviously fixed to a simple plateau profile. The plateau became rounded. For some stimulation sites, complex patterns were observed in which the agonist muscle became active in an initial burst followed at longer latency by a rise in activity in the antagonist muscle. Such biphasic patterns are typical of normal movement. However, for most sites, the overall trend was toward a simple plateau of activity that was maintained during the stimulation train. Thus only a hint of normal EMG patterning was observed. It is possible that a stimulation train that is modulated in a more normal temporal pattern, with rising and falling activity matched to the activity of motor cortex neurons, might more successfully mimic the normal temporal pattern of muscle activity.

In a third experiment, we tested whether the evoked muscle activity depended on the initial position of the limb (Graziano, Patel, et al., 2004). If motor cortex operates through a fixed connection to the muscles, with a relay in the spinal cord, then stimulation of the same cortical site should always result in the same levels of muscle activity regardless of the position of the limb. Such a result is unlikely because Sanes et al. (1992) already showed in the rat motor cortex that the pattern of evoked muscle activity in the limb depends on the position of the limb. We anesthetized the animal and fixed the limb in a holder. The elbow joint could be held at four different angles. Figure 7-8A shows the result for one stimulation site. Muscle activity rose on stimulation onset, remained at an approximate plateau during the stimulation train, and fell back to baseline at stimulation offset. When the elbow was fixed in

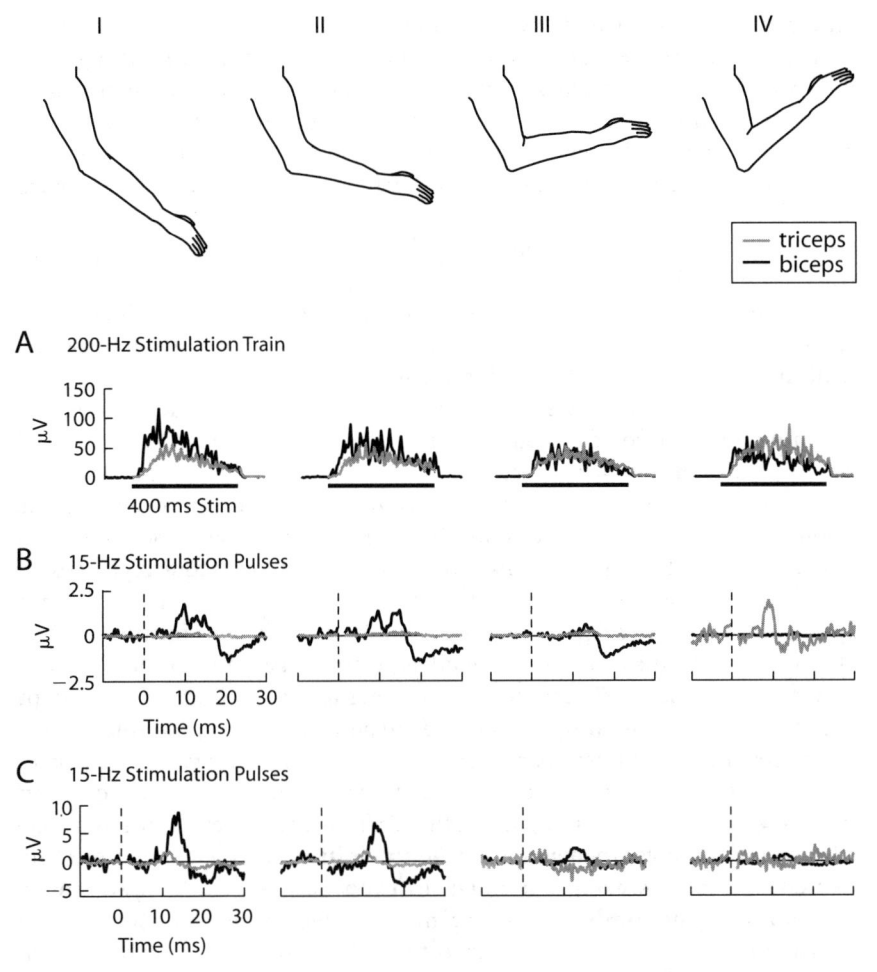

7. Complex Movements Evoked by Electrical Stimulation

Figure 7-8 Cortico-muscle connectivity modulated by proprioceptive feedback. Top: The arm was fixed in four possible locations in an anesthetized monkey. **A.** Muscle activity in triceps and biceps evoked by stimulation of one site in primary motor cortex. Stimulation trains were presented for 400 ms at 200 Hz. Mean of 20 stimulation trains. **B.** Muscle activity in triceps and biceps evoked by stimulation of the same site in primary motor cortex using stimulation pulses presented at 15 Hz. Vertical line on each histogram indicates time of stimulation pulse delivered to brain. Time from 0.2 ms before to 1.5 ms after the pulse was removed from the data to avoid electrical artifact. Each histogram is a mean of 2000–4500 pulses. Stimulation of this point in cortex could activate the biceps or the triceps depending on the angle of the joint, consistent with driving the elbow toward an intermediate angle. **C.** A second example site in primary motor cortex. Stimulation of this site in cortex activated primarily the biceps. When the elbow was far from a flexed position, stimulation evoked a higher level of biceps activity and a greater discrepancy between biceps and triceps activity, consistent with driving the elbow in a regulated fashion toward flexion. Adapted from Graziano, Patel, et al. (2004).

different angles, the activity level evoked in the biceps and triceps changed. Moreover the relative signal strength between these two muscles changed. In this case, when the elbow was extended, stimulation evoked high biceps activity relative to triceps activity, as though to drive the arm toward a more flexed posture. When the elbow was flexed, stimulation evoked lower biceps activity relative to triceps activity, as though to drive the arm toward a more extended posture. In this manner stimulation evoked the appropriate spatial pattern of muscle activity to drive the elbow toward a goal angle. This specific pattern of result is essentially a replication of the result of Sanes et al. (1992) in the rat.

In a final study, rather than testing the average effect of a long train of stimulation pulses, we examined the average effect of a single stimulation pulse. We used the technique of stimulus triggered averages (Cheney et al., 1985). In an anesthetized monkey, pulses of current were delivered to motor cortex, the interpulse intervals long enough (66 ms) that the pulses could be considered approximately independent events.

Figure 7-8B shows the result for one example site in motor cortex. The elbow was fixed at four possible angles while the stimulus pulses were presented to cortex. Because the pulses were presented at low frequency, they had no visible effect on the arm. The arm did not appear to strain against the arm holder. Yet when averaged over 3000 pulses, a minute effect at the muscle emerged. When the elbow was fixed in an extended posture, stimulation evoked a short latency (approximately 7 ms) increase in biceps activity but little or no triceps activity, consistent with a signal to flex the elbow. In contrast, when the elbow was fixed in a flexed posture, stimulation evoked a short latency increase in triceps activity but little or no biceps activity, consistent with a signal to extend the elbow. The effect of stimulating this cortical site therefore had opposite effects on the muscles depending on the angle at which the elbow was fixed. The effect was consistent with a signal to move the elbow from any initial position to an intermediate final angle. Just such a movement was obtained with a long stimulation train at that cortical site.

Figure 7-8C shows the result for another example site in motor cortex. When the elbow was fixed in an extended posture, stimulation evoked a short latency increase in biceps activity and little or no triceps activity, consistent with a signal to flex the elbow. When the elbow was fixed in a flexed posture, stimulation evoked little activity in either muscle. This pattern of activity is consistent with a signal to move the elbow toward flexion in a regulated manner, such that when the elbow is far from a flexed posture, the muscle force is larger, and when the elbow is already at a flexed posture, the muscle force is minimal. Just such a movement toward elbow flexion was obtained with a long stimulation train at that cortical site.

These studies on EMG patterns during stimulation show that the cortical stimulation evokes some features of normal movement but not an exact replica. The spatial pattern of activity across muscles changes depending on the starting position of the limb, consistent with pulling the limb in the correct direction toward the goal posture. The temporal pattern of activity shows less of a natural profile, with stimulation of many sites evoking a squared temporal profile that follows the time course of the stimulation train. These effects on muscle activity, the similarities and differences to natural movement, are further discussed in Chapter 11 in which the possible cortico-spinal pathways and neuronal mechanisms are considered.

RESULTS 2: QUALITATIVE DESCRIPTION OF ACTION CATEGORIES

Seven categories of movement and the approximate cortical zones from which they could be evoked are shown in Figure 7-9.

Hand-to-Mouth Movements

Stimulation within a restricted zone in the precentral gyrus evoked a characteristic hand-to-mouth movement. Five components were typical of this movement. The grip aperture closed in the hand contralateral to the electrode; the

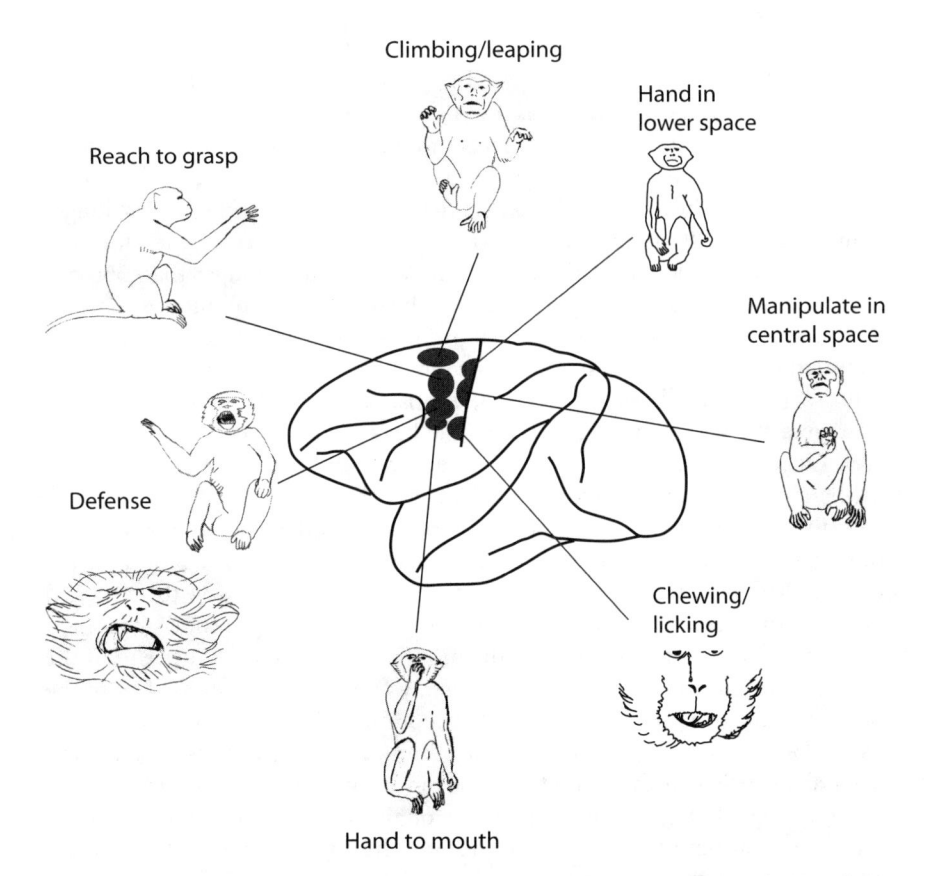

Figure 7-9 Common categories of movement evoked by electrical stimulation of the motor cortex in monkeys, using the behaviorally relevant time scale of 0.5 sec. Images traced from video frames. Each image represents the final posture obtained at the end of the stimulation-evoked movement.

forearm supinated and the wrist flexed, such that the grip was aimed at the mouth; the elbow flexed and the shoulder rotated such that the hand moved precisely to the mouth; the mouth opened; when the head was released from the headbolt and allowed to turn freely, stimulation caused a rotation of the head to a forward-facing position, contributing to the alignment of the mouth and the hand. These five movement components occurred simultaneously in a coordinated fashion resembling the monkey's own voluntary hand-to-mouth movements.

Although the movements resembled voluntary actions in some respects, they clearly were not true voluntary movements of the monkey's but were driven by the stimulation. Typically, the movement could be obtained on every stimulation at short latency with mechanical reliability for hundreds of trials, with no adaptation or degradation. Similar movements could be evoked in anesthetized animals, though the movements were weaker and required greater current under anesthesia. A short stimulation, such as a 100-ms stimulation, evoked the initial part of the action, a slight closing of the hand, a slight twitching of the hand upward in the direction of the face, and a slight opening of the mouth. This truncated movement, by itself, makes no behavioral sense. It is best described as a twitch. It makes sense, however, if interpreted as the initial segment of a larger movement that has not had time to unfold. Longer stimulations, such as for 300 ms, allowed more of the movement to unfold, but rarely allowed the hand to reach the mouth. Yet longer stimulation of 500 ms almost always allowed the hand to reach the mouth in an apparent completion of the movement. Stimulations longer than 500 ms, such as those of 1000 ms, typically caused the hand, arm, and mouth to freeze at the final configuration, as if the movement had been completed and the activated circuit were maintaining the final posture. When the stimulation train was extended beyond 1 second, almost always the animal appeared to overcome the stimulation effects and take back some degree of control of its arm. Once the stimulation train stopped, however, and then was reinitiated, the hand would move directly back to the mouth.

If the monkey was reaching toward a piece of food at the time of stimulation onset, the hand would close on empty air and come to the mouth. If the monkey had just grasped a piece of food, stimulation would drive the clenched hand to the mouth and cause the hand to freeze at the mouth, the food securely gripped in the fingers and the mouth stuck open, until the end of the stimulation train, at which time the animal would finally be released from the stimulation-evoked posture and put the food in its mouth. If an obstacle was placed between the hand and the mouth, stimulation would cause the hand to move along a direct path toward the mouth and bump against the obstacle, pressing against it throughout the remainder of the stimulation, without moving intelligently around the obstacle. Therefore, although the stimulation evoked a movement of great complexity and coordination, the complexity was also limited. The movement resembled a fragment of behavior that was mechanically produced by the stimulated circuitry without intelligent flexibility.

Not all sites within the hand-to-mouth zone resulted in the same movement. For example, depending on the cortical site, stimulation drove the hand to one side of the mouth or the other, and caused the mouth to open more on the side that the hand approached, as if the monkey were placing a piece of food into the side of the jaw, as the animals often do in normal behavior. Not only did the exact position of the hand vary from site to site, but the type of hand grip also varied. For some stimulation sites the hand shaped into an apparent precision grip, the thumb against the side of the forefinger (typical of a macaque precision grip). For other stimulation sites, the hand shaped into what we called a hamburger grip, the four fingers against each other and opposed to the thumb, with a gap between, as if for gripping a larger object. These variations suggested that the zone of cortex was not uniform and not dedicated to producing a single movement, but instead probably contributed to a range of movements that fell within the large class of interactions between the hand and the mouth. In normal monkey behavior, the hand is often brought to the mouth to put in food, take out food, manipulate a piece of food that is in the mouth, scratch the lips, pick at the teeth, push food out of the cheek pouches, and so on.

It is unlikely that the collection of components in a hand-to-mouth movement co-occurred by chance. Even putting aside the specific combination of body parts, the hand closes rather than opens (50% chance); the mouth opens rather than closes (50% chance); the forearm supinates, aiming the grip at the mouth, rather than pronates, aiming the grip away from the mouth (50% chance); the hand moves within about 5 cm of the mouth, a ball of space accounting for about 1% of the total workspace of the hand (1% chance); and the head turns to a forward position, within about 5% of its range of motion (5% chance). Multiplied, these conservatively estimated probabilities yield a p value of 0.00005. We must dispense with the occasionally suggested interpretation that the evoked movements are chance collections of twitches rather than meaningful fragments of the behavioral repertoire.

In all monkeys tested, the hand-to-mouth sites were clustered in a lateral, anterior zone probably within the ventral premotor cortex. Whether they are in the caudal or rostral division is unclear. Every monkey tested had a hand-to-mouth zone, but the exact location varied somewhat, especially in the rostrocaudal dimension. Our current interpretation is that the hand-to-mouth sites are more likely to lie within a ventral anterior part of F4 as defined by Matelli et al. (1985) and that the dorsal part of F4 emphasizes a different type of action, the defensive movement.

Defensive Movements

In a specific zone in the precentral gyrus, neurons typically respond to tactile stimuli on the face and arms and to visual stimuli looming toward the tactile receptive fields (Fogassi et al., 1996; Gentilucci et al., 1988; Graziano et al., 1997a; Rizzolatti et al., 1981). Some of the neurons are trimodal, responding also to auditory stimuli in the space near their tactile receptive fields (Graziano et al., 1999). Because of these distinctive sensory properties, we refer to this cortical region as the polysensory zone (PZ). Although all monkeys tested have a PZ, it varies among animals in size and precise position (Graziano and Gandhi, 2000). It is typically located just posterior to the bend in the arcuate sulcus. In the terminology scheme of Matelli et al. (1985), it probably corresponds to the dorsal part of premotor area F4 where similar polysensory neurons have been reported (Fogassi et al., 1996; Gentilucci et al., 1988).

Stimulation within this zone evokes movements that closely resemble a natural defense of the body surface such as to an impending impact or unexpected touch. For example, at some sites, the neurons had tactile receptive fields on the side of the face contralateral to the electrode and visual receptive fields in the space near that side of the face. Stimulation of these sites evoked a defensive action that included seven components: a blink, stronger or exclusively on the contralateral side; a squinting of the musculature surrounding the eye; a lifting of the upper lip in a facial grimace that wrinkled the cheek upward toward the eye; a folding of the contralateral ear against the side of the head; a shrugging of the shoulder, either stronger on or exclusively on the contralateral side; a rapid turning of the head away from the contralateral side; a rapid lifting of the arm, sweeping the hand and forearm into the contralateral space near the face as if blocking or wiping away a potential threat; and a centering movement of the eyes (Cooke and Graziano, 2004a; Graziano, Taylor, et al., 2002). These movement components match point for point the components of a normal defensive reaction such as when the monkey's face is puffed with air (Cooke and Graziano, 2003).

At other sites, neurons had a tactile receptive field on the arm and hand and a visual response to objects looming toward the arm and hand. Stimulation caused a fast retraction of the hand to the side or back of the torso. In general, the movement evoked from a site within PZ seemed appropriate for defending the part of the body covered by the tactile and visual receptive fields of the neurons.

We observed apparent summation between the stimulation-evoked defensive-like movements and actual defensive movements. In the summation test, we lowered the stimulating current to a point near or below threshold until a subtle movement was obtained only on some trials. We then puffed air on the monkey's face, or presented some other noxious stimulus such as a ping pong ball thrown at the animal, evoking a defensive reaction. Within a second after the actual defensive reaction, we then stimulated the site in PZ. Under this condition, the stimulation evoked a robust, superthreshold defensive reaction. The actual defensive movement seemed to prime the system such that a low stimulating current in PZ could evoke a large effect.

One possibility is that the stimulation of sites in PZ evoked a noxious sensory percept to which the monkey then reacted. This possibility is difficult or impossible to rule out because the monkey cannot self-report. However some observations suggest that it is unlikely. Although the stimulation evoked an apparent defensive reaction, as soon as the stimulation train ended the reaction

7. Complex Movements Evoked by Electrical Stimulation

ended and the monkey returned to feeding itself or playing with toys. A brief stimulation, such as for 50 ms, evoked a correspondingly brief movement, shorter than any behaviorally normal defensive reaction; a long stimulation, such as for 1000 ms, evoked a correspondingly sustained movement that terminated abruptly at the end of the stimulation. An actual noxious stimulus, such as an air puff or a ping pong ball thrown at the face, did not result in such tight time-locking to the stimulus, but instead resulted in an extended reaction including general agitation and threats to the experimenter. Moreover, the defensive-like movements evoked by stimulation could still be evoked under anesthesia, even when the anesthesia was so deep that the animal did not react to noxious stimuli.

To further test the role of PZ in the coordination of defensive movements, we disinhibited neuronal activity in PZ by injecting the chemical *bicuculline* and inhibited neuronal activity by injecting the chemical *muscimol* (Cooke and Graziano, 2004b).

When bicuculine was injected into PZ, not only did the local neuronal activity increase, but the neurons also began to fire in intense spontaneous bursts of activity with approximately 5 to 30 sec between bursts. Each spontaneous burst of neuronal activity was followed at short latency by the standard set of defensive-like movements, including blinking, squinting, flattening the ear against the side of the head, elevating the upper lip, shifting the head away from the sensory receptive fields, shrugging the shoulder, rapidly lifting the hand into the space near the side of the head as if to block an impending impact, and centering the gaze. Chemical stimulation of neurons within PZ, therefore, produced the same effect as electrical stimulation. This result may seem expected. If electrical stimulation of PZ evokes a set of movements, then surely chemical stimulation should too. However, chemical stimulation is in some ways a more specific manipulation, affecting local neuronal receptors. It does not stimulate fibers of passage or induce antidromic activation. The result of chemical stimulation in PZ, therefore, is an important confirmation and strengthens the findings from electrical stimulation.

In addition to evoking defensive-like movements by inducing bursts of neuronal activity, bicuculline also altered the monkey's actual defensive reaction to an air puff directed at the face. After the injection of bicuculline into PZ, the monkey gave an exaggerated defensive reaction to the air puff. The magnitude of the defensive reaction, as measured by facial muscle activity, was approximately 45% larger after bicuculline injection than before injection. Even gently bringing a Q-tip toward the face, normally evoking little reaction from the monkey, evoked a pronounced defensive reaction in the monkey with a bicuculline-treated PZ. Muscle activity during chewing, threat faces, and eyebrow movement was not elevated. The effect was limited to the defensive reaction.

When muscimol was injected into PZ, thereby inhibiting neuronal activity, the monkey's defensive reaction to the air puff was reduced. The magnitude of the defensive reaction, as measured by facial muscle activity, was approximately 30% smaller after muscimol injection than before injection. Injections into surrounding cortical tissue outside of PZ did not affect the defensive response to an air puff. These chemical manipulations therefore strengthen the case for PZ as a sensory-motor interface related to the defense of the body surface, a cortical region to which the appropriate visual, tactile, and auditory information is supplied, and from which emerges the motor command to produce spatially directed defensive reactions.

One more observation about PZ is worth noting. In an experiment that we conducted before we understood the possible link between PZ and defensive movements, we observed that a typical multimodal neuron in PZ would respond robustly to the sight of a rubber snake near the monkey's body, in the visual receptive field of the neuron (Graziano, Alisharan, et al., 2002). In contrast, the same neuron would respond weakly or not at all to the sight of an apple in the same region of space near the body, though the monkey appeared to be equally interested in both stimuli. At the time, not understanding the implication of this bizarre but consistent observation, we facetiously called the neurons "biblical cells" and did not attempt to explain them. In retrospect, it seems the neurons in PZ were especially responsive to threatening stimuli.

Manipulation Movements

Stimulation of another cluster of sites evoked an especially varied and complex set of movements that involved the fingers, wrist, and often the arm and shoulder, contralateral to the electrode. The movements resembled the types of actions that monkeys typically make when manipulating, examining, or tearing objects. The finger movements included an apparent precision grip (thumb against forefinger), a power grip (fist), or a splaying of the fingers. In some cases a supination or pronation of the forearm occurred, rotating the grip one direction or the other. Also in some cases the wrist flexed or extended. A common action for monkeys is to splay the fingers of one hand, orient the palm toward the face, and examine the splayed hand, perhaps searching for stray granules of food. This splayed-hand posture, with the palm oriented toward the face, was often evoked on stimulation within this cortical zone. Monkeys commonly manipulate objects in a region of central space within about 10 cm of the chest. Stimulation within this cortical zone often evoked a movement of the shoulder and arm that brought the hand into this central region of space. A common action for monkeys when manipulating objects is to tear the object or pull it in two, the two hands pulling rapidly from central space toward lateral space while the forearms supinate and the hands are tightly gripped. Stimulation within this zone of cortex also sometimes evoked just such a movement, though only in the contralateral limb.

These sites were clustered in a posterior zone that lay partly on the gyral surface and partly on the anterior bank of the central sulcus. This cluster probably corresponds to the traditional primary motor hand representation. It may also correspond to the central hand region in the motor cortex maps of Kwan et al. (1978) and Park et al. (2001). We suggested that this cortical zone may represent a "manual fovea," a repertoire of movements that is related to the manipulation of objects and that is heavily biased toward but not exclusively limited to hand locations in a central region of space in front of the chest (Graziano, Cooke, et al., 2004).

Reach-to-Grasp Movements

For some cortical sites stimulation evoked an apparent reach in which the wrist straightened, the fingers opened as if to grasp, the forearm pronated to orient the grip outward, and the hand extended away from the body. In some cases the hand extended to a region of space as far as 25 cm distant from the body, with the arm straight. In other cases the hand converged on a location at a lesser distance, with the elbow partially flexed, as if the hand were reaching to a closer object. In all of these cases stimulation caused a convergence to the final posture from a range of initial postures. These apparent reaching sites tended to be located on the gyral surface just anterior to the "central space/manipulation" zone and dorsal to the "defensive" zone. Because of this relative location, the reach-related sites probably lie within the dorsal premotor cortex, within the PMDc, where a high proportion of neurons respond in relation to reaching movements (e.g., Crammond and Kalaska, 1996; Hocherman and Wise, 1991; Johnson et al., 1996; Messier and Kalaska, 2000). Typically stimulation of more rostral sites did not evoke reliable or clear movements.

Hand in Lower Space

A commonly evoked movement involved a placement of the hand in lower space near the feet, typically with the forearm pronated such that the palm faced down or inward toward the body. These stimulation-evoked movements resembled a common part of the monkey's behavioral repertoire in which the hand was braced on the ground (Graziano, Cooke, et al., 2004). These sites were typically found just dorsal to the central space/manipulation sites.

Mouth Movements

The above movement categories were evoked from the large arm and hand sector of the lateral motor cortex. When we stimulated in cortex ventral to the arm and hand representation, we obtained movement of the jaw, lips, and tongue, as expected on the basis of the standard body map described for the motor cortex (Woolsey et al., 1952). The mouth movements often appeared to be coordinated and of behavioral significance. For example, stimulation of one site caused the jaw to attain a partially open position, the lips to purse slightly toward the contralateral side of the mouth, and the tongue to move until the tip was placed in a contralateral and slightly protuberant position. The final oral posture evoked from this site resembled an action to acquire a bit of food just outside the mouth on the contralateral side. We looked for but did not find any obvious cortical map in the mouth representation in terms of the spatial location around the mouth toward which the tongue and lip movement was directed. We also did not obtain movements that looked like threat displays, fear grimaces, or any other social displays. It is likely, however, that we failed to discover many of the movement types in the mouth representation because we did not explore it as extensively as we did the arm and hand representation. We rarely obtained rhythmic chewing-like jaw movements, perhaps because we used stimulation trains of 0.5 sec in duration instead of the 3-sec train durations used by Huang et al. (1989) who reported rhythmic chewing movements.

Climbing/Leaping

In a medial and anterior region, stimulation evoked especially complex movements that involved bilateral action of the arm and leg, movements of the torso, and movements of the tail, often simultaneously from one site. These complex, whole-body sites correspond roughly to the SMA proper, a cortical region on the crown of the hemisphere and extending slightly onto the lateral side, just anterior to the primary motor leg representation. Others have also obtained bilateral movements of multiple body parts on stimulating in this area of cortex (e.g., Foerster, 1936; Luppino et al., 1991; Penfield and Welch, 1951; Woolsey et al., 1952).

Subjectively, the movements resembled climbing or leaping postures. For example, stimulation of one site caused the left foot to press down against the floor of the primate chair, the right foot to lift and reach forward with the toes shaped as if in preparation to grasp, the left hand to reach toward a lower, lateral position while shaped as if in preparation to grasp, the right hand to reach toward a position above the head while shaped as if in preparation to grasp, and the tail to curl to one side. The long-tailed macaques in our experiments do not have prehensile tails. They use their long, stiff tails mainly as balance devices during locomotion, and therefore the tail movements evoked by stimulation of SMA are consistent with a possible role in locomotion.

Stimulation within the SMA did not always evoke bilateral movements. For example, stimulation of another site caused the lower torso to turn to the left side, the left foot to reach out and down as if stepping to a position lateral to and slightly behind the body, and the left arm to reach to a lateral position as if to grasp a support.

Although we sometimes tested stimulations extended to 1 sec or more, we did not observe any cyclical stepping movements. Instead the movements resembled the complex adjustments of body and limb often seen when monkeys are navigating a complex environment. The climbing-like movements, however, were restricted by the primate chair in which the animal was tested and therefore could never be compared directly to the normal climbing, leaping, or complex locomotor movements of a monkey.

COMPLEX MOVEMENTS REPORTED IN OTHER STUDIES

Midbrain

In a now-classic study, electrical stimulation of a midbrain nucleus in the cat resulted in patterned locomotor behavior (Shik et al., 1969). The exact role of this mesencephalic locomotor nucleus, its relationship to spinal and cortical control of locomotion, is still unknown.

Electrical stimulation has long been used to study maps of motor output in the superior colliculus or, as it is called in nonmammals, the optic tectum. The map of saccadic eye movements in cats and monkeys is perhaps the best-known result in the colliculus (Guitton et al., 1980; Robinson, 1972; Schiller and Stryker, 1972). However, other complex species-typical behaviors can be evoked. Stimulation of the optic tectum in salamanders evokes a coordinated movement in which the animal orients to a spatial location, reaches out with the forepaws, and opens the mouth as if to acquire prey (Finkenstadt and Ewert, 1983). In rats, stimulation of the part of the map that represents lower visual space evokes orienting movements of the head as if the animal were acquiring an object on the ground in front of it, and stimulation of the part of the map that represents upper visual space evokes retracting, defensive-like movements (Dean et al., 1989). These movements are consistent with the exigencies of normal life for a rat, in which food is found on the ground in lower visual space and enemies attack from above.

Spinal Cord

Giszter et al. (1993) electrically stimulated sites in the spinal cord of frogs and studied the effect on the hind leg. The frog's ankle was fixed in a range of different spatial locations. For each ankle location, the force evoked by stimulation was measured. These stimulation-evoked forces formed a convergent force field pointing toward a single location in space, suggesting that if the ankle were free to move, the foot would move to that spatial terminus. Different stimulation sites resulted in convergent force fields aimed at different spatial locations.

Our results on stimulating the monkey motor cortex are similar in that stimulation caused the limb to converge from a range of initial locations toward a specific final location. Presumably the cortical stimulation operates by recruiting spinal circuitry. If the Giszter et al. (1993) result is applicable to the monkey spinal cord, then our results may depend on spinally mediated force fields. The cortical stimulation, however, appears to recruit a higher-order or more integrated version of the spinal force fields. We typically found convergence of many joints from different body segments. For example, a hand-to-mouth movement involves a coordination of output that passes through the spinal cord (for the control of the hand, arm, and shoulder) and through the facial nucleus (for the control of the mouth). The possible relationship between the movements evoked from spinal stimulation and the movements evoked in our studies of motor cortex is discussed in greater detail in Chapter 11.

Cortex

Several studies have confirmed the essential phenomenon of complex movements evoked from the motor cortex and have now extended the results to other species of animals.

In the rat, species-typical behavior can be evoked by stimulation of motor cortex. Brecht et al. (2004) found that intracellular stimulation of a single cortical neuron evoked rhythmic whisking movements. Cramer and Keller (2006) suggested that the cortically controlled whisking actions are mediated by a projection from the motor cortex to a subcortical central pattern generator that in turn controls the whiskers.

Haiss and Schwarz (2005) found that the rat motor cortex contained two adjacent zones related to the whiskers. Stimulation of one zone on a behavioral time scale (500 ms) evoked rhythmic whisking similar to normal exploratory movements. Stimulation of the other zone for 500 ms evoked a retraction of the whiskers on the contralateral side, a closure of the contralateral eye, a facial grimace, and sometimes a lifting of the contralateral forepaw to the space beside the face. These results suggest that the rat motor cortex, like the monkey motor cortex, may be organized into zones that emphasize different ethologically useful actions, in this case exploratory whisking for one zone and defensive actions for the other zone. Haiss and Schwarz (2005) also emphasize the important point that these two zones might not act as separate modules but instead might act in concert to organize complex behavior.

Ramanathan et al. (2006) stimulated the rat motor cortex on a behaviorally relevant time scale (500 ms) and obtained reaching and grasping movements of the forepaws. Moreover, when a reaching zone in cortex was lesioned, the rat's ability to reach was compromised. When the rat was retrained to reach, the motor cortex was found to have reorganized such that reaching could be electrically evoked from new cortical sites. After this rehabilitation, the ability of the rat to perform the behavior correlated with the amount of cortex that, when stimulated, evoked the behavior. These results strongly support the view that the motor cortex is organized to control complex, meaningful behavior, that different behaviors are emphasized in different regions of cortex, and that these behaviors can be assessed through electrical stimulation.

In the cat motor cortex, Ethier et al. (2006) found that stimulation on a behaviorally relevant time scale (500 ms) evoked a variety of forepaw movements including apparent reaching and hooking of the paw as if to acquire an object.

Stepniewska et al. (2005) stimulated the motor and posterior parietal cortex of galagos using the behaviorally relevant time scale of 500 ms. They evoked complex movements that resembled fragments of the animal's normal behavioral repertoire. Different categories of movement were evoked from different cortical zones. The posterior parietal lobe could be segmented into functional zones including a hand-to-mouth zone, a defensive zone, and a reaching zone. Similar results were obtained in the motor cortex, but were not studied in as much detail.

In macaque monkeys, stimulation of the ventral intraparietal area (VIP) evokes movements that resemble defensive reactions (Cooke and Graziano, 2004a; Cooke et al., 2003; Graziano, Taylor, et al., 2002b; Thier and Andersen, 1998). The evoked movements are similar to those evoked from PZ in the motor cortex. VIP is anatomically connected to PZ (Lewis and Van Essen, 2000; Luppino et al., 1999), and neurons in VIP respond preferentially to visual stimuli looming toward the face, auditory stimuli near the face, and tactile stimuli on the face, much like neurons in PZ (e.g., Colby et al., 1993; Duhamel et al., 1998; Schaafsma and Duysens, 1996; Schlack et al., 2005; Zhang et al., 2004). One possibility is that VIP and PZ are part of a larger circuit that contributes to the maintenance of a margin of safety around the body (Graziano and Cooke, 2006). A second suggestion is that VIP contributes to navigation with respect to nearby objects (Schlack et al., 2005). These two suggestions are complimentary because navigation with respect to nearby objects is mainly a process of obstacle avoidance.

Two general lessons emerge from these many studies. First, electrical stimulation applied on a behavioral time scale is a useful way to study motor areas. It provides an initial description of function. The hypotheses generated by electrical stimulation can then be tested in greater detail using other techniques. Second, motor areas tend to be organized around actions that are of ethological importance to the animal. To study the motor system, therefore, it is essential to understand the behavioral repertoire of the species under study.

Chapter 8

The Match Between Natural Neuronal Properties and Stimulation-Evoked Movement

INTRODUCTION

When monkeys are trained to reach from a central location to a fixed set of targets, neurons in motor cortex are tuned to the direction that the hand moves (e.g., Georgopoulos et al., 1982; Georgopoulos et al., 1986; Georgopoulos et al., 1988; Kettner et al., 1988; Schwartz et al., 1988). Typically a neuron will become active during movement of the hand in one preferred direction and will become less active during movement of the hand in other directions. In our electrical stimulation studies, however, stimulation of a site in cortex did not evoke a specific direction of hand movement. Instead the stimulation tended to drive the arm toward a specific final postural configuration. One consequence of the movement of the arm to a goal posture was a movement of the hand from any initial position to a goal region of space. Stimulation could evoke any direction of hand movement, depending on the initial position of the hand relative to the goal region. This profound mismatch between the single-neuron properties and the stimulation properties is problematical. In other brain areas, as reviewed in Chapter 6, when electrical stimulation was applied with the same durations and currents that we used in motor cortex, the results closely matched the properties of single neurons at the stimulated site. Why does a mismatch exist in motor cortex?

This mismatch between neuronal properties and stimulation-evoked movement appears to stem from a set of ambiguities in interpreting the single-neuron findings. Because the single-neuron technique is correlative, it is interpretationally murky. Correlations can change in type and magnitude depending on the exact design of the study and the focus of the analysis. We explicitly tested how the tuning of neurons to different parameters can change radically depending on the movement set that is selected and on the method of analysis (Aflalo and Graziano, 2006a, 2007). In the process, we tracked down the reasons for the apparent mismatch between single-neuron properties and stimulation effects. There are two main reasons.

First, most previous studies of the tuning properties of motor cortex neurons explored the direction of the hand through a full, 360-degree range of angles, spanning a circle or a sphere. Each neuron was broadly tuned to direction, and the tuning curve emerged only over this full range of angles. Testing over a small wedge of angles would have uncovered a jagged, local piece of the larger tuning curve and therefore would not have revealed any systematic direction tuning. The movement of the hand to and from different locations in space, however, was tested in restricted regions of space approximating 5% of the normal range. Even those studies that tested an expanded spatial range encompassed about 15% of the hand's workspace or less (e.g., Caminiti et al., 1990; Fu et al., 1993; Kettner et al., 1988; Sergio and Kalaska, 2003). By testing direction over 100% of its range and position over 5% to 15% of its range, previous studies arrived at the conclusion that neurons were tuned mainly to direction and little if at all to the goal position of the movement.

A second reason for the apparent mismatch between single-neuron findings and stimulation results is that many previous studies examined the tuning of neurons to hand movement without examining the multijoint posture of the arm. The primary effect of stimulation is to drive the arm joints toward a specific posture, such as in the case of a hand-to-mouth movement. Consider a hypothetical neuron that fires best during a hand-to-mouth movement, and that is tested in a directional reaching task. One direction of reach may be aimed up and toward the mouth. But the posture is wrong. The palm faces away from the mouth, the grip is open, and the wrist may be extended. Although the hand moves toward the right location, the arm does not move toward the right posture, and therefore the tuning of this neuron is obscured. The point is that one cannot adequately address a neuronal tuning to posture in an experiment that manipulates and tracks only hand position. The richness and complexity of posture space, which is at least eight dimensional with respect to the arm, cannot be captured with the three dimensions of hand position space.

Once these considerations were taken into account, we were able to demonstrate a direct match between the movements evoked by stimulation of a site in cortex and the properties of single neurons at that same site. Stimulation drives the arm to a particular posture. During the monkey's own spontaneous movement, the neurons at that same cortical site tend to fire at a high rate during movements that terminate near the same posture. One function of the neurons at a cortical site therefore appears to be to cause the arm to move from any initial configuration to a particular final posture. Our interpretation is that the true function of the neurons at a cortical site is to cause the arm to make a behaviorally useful movement, and the movement repertoire of monkeys is dominated by behaviorally useful postures.

In the experiment described in this chapter, the monkey sat in a primate chair with one arm free to move spontaneously and naturalistically over the full range of the workspace. The monkey touched and explored parts of the primate chair, reached for pieces of fruit held out on the end of forceps in a diversity of locations in the workspace of the arm, brought the food to the mouth, retrieved food from the mouth, held and examined food in central space, rotated and explored objects, scratched at its fur, scratched rhythmically at portions of the monkey chair, and attempted to scratch the experimenter with fast semiballistic arm movements. The configuration of the arm and the grip aperture, totaling eight degrees of freedom, were measured at high spatial and temporal resolution by means of lights fixed to points on the arm and hand. At the same time, the activity of neurons in motor cortex was measured. In this paradigm, a great range of movement parameters were in play. The goal of the analysis was to correlate these movement parameters with neuronal activity.

In the following sections only a subset of the analyses are described, focusing on the tuning of neurons to hand direction and arm posture. A more extensive set of analyses can be found in Aflalo and Graziano (2007). The results of the analyses indicate a three-way match between the properties of single neurons in motor cortex, the effects of electrical stimulation in motor cortex, and the behavioral repertoire of monkeys. Neurons in motor cortex are tuned to many movement parameters with an emphasis on the goal posture of the arm; electrical stimulation evokes complex movements for which the final posture of the arm is the most consistent feature; and the behavioral repertoire of a monkey emphasizes useful arm postures that are maintained within relatively narrow tolerance during the performance of common actions.

DIRECTION TUNING: LOCAL AND GLOBAL

The first question addressed was whether the classical finding of direction tuning in motor cortex neurons (e.g., Georgopoulos et al., 1982; Georgopoulos et al., 1986; Georgopoulos et al., 1988; Schwartz et al., 1988) could be replicated using the free movement paradigm of our experiment. An analysis for one example neuron is shown in Figure 8-1. The movement of the arm was divided into separate hand movements on the basis of the classical bell-shaped pattern of a rising and falling hand speed. During the testing of this neuron, 320 hand movements could be distinguished. Of this full set of 320 movements, we selected a subset that resembled the center-out reaching task. In this subset, the hand trajectories began within a restricted central region of space 5 cm in radius, and extended outward between 6 cm and 15 cm. This subset is shown in Figure 8-1A.

We performed an analysis using this subset of movements to determine if the neuron was significantly tuned to hand direction. First each hand movement was assigned a direction, defined as the vector connecting the start location of the hand trajectory to the end location. Then the mean firing rate of the neuron during each movement was calculated. Finally these directions and associated firing rates were fitted to a tuning function using a regression analysis. We used a standard cosine tuning model (Georgopoulos et al., 1986). In this tuning model, a neuron has a preferred direction. The neuron is most active during movements in which the hand direction is close to the preferred direction, with a small angular difference ($\Delta\theta$). The neuron is least active during movements in which the hand direction is far from the preferred direction, with a large $\Delta\theta$.

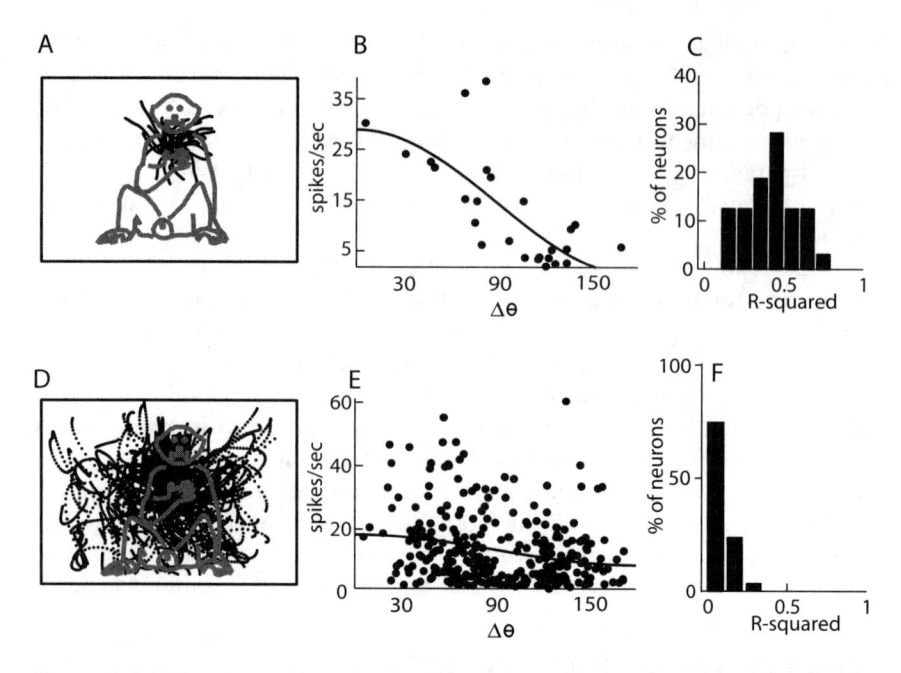

Figure 8-1: Direction tuning of motor cortex neurons. **A.** Front view of 26 selected hand movements made during 10 min of testing one neuron. Each trail of dots=1 movement measured at 14.3 ms intervals. Frame is 45 cm tall. Each movement shown originated within a 5-cm radius sphere of central space and extended any direction in three-dimensional Cartesian space for a distance between 6 cm and 15 cm. **B.** Tuning of an example neuron to direction, based on selected movement set. X axis shows angular difference between the direction of each movement and the preferred direction; Y axis shows mean firing rate during each movement; for cosine tuning to direction, $R^2=0.43$, $p=0.0001$. **C.** Frequency histogram of R^2 values for all neurons tested as in B. **D.** Front view of full set of 320 hand movements made during testing of one neuron. **E.** Direction tuning of an example neuron (same neuron as in B), based on full movement set. $R^2=0.05$, $p=0.00008$. Note that a new preferred direction was obtained by regression, and therefore the data points shown in B do not plot to the same location on the x axis as in E. **F.** Frequency histogram of R^2 values for all neurons tested as in E. Adapted from Aflalo and Graziano (2007).

Figure 8-1B shows the tuning curve for this neuron. The regression analysis was able to find a preferred direction that fit the data. On average, the firing rate was high during movements near the preferred direction (thus with low $\Delta\theta$) and low during movements far from the preferred direction (with high $\Delta\theta$). The cosine tuning is shown as a function line through the data points. The regression analysis returned an R^2 value of 0.43. This R^2 value indicates that 43% of the neuronal variance could be explained by the model of direction tuning. The fit to the model was highly significant ($p=0.0001$).

Figure 8-1C shows the R^2 values for all sixty-four cells tested. The mean R^2 value was 0.42, and 68% of the cells showed a significant fit to the cosine model of directional tuning.

The present results are therefore similar to previous results using a center-out task. Most neurons showed a significant fit to a cosine tuning function, with nearly half of the neuronal variance attributable to direction tuning. Even with unrestricted, untrained movements, once a center-out subset of movements is selected, the tuning properties of the neurons closely resemble previous accounts.

Figure 8-1D shows the full set of 320 movements for the same example neuron. In this global movement set, there is no selection of central movements. All movements, regardless of length, start position, or end position, are included. As a result, hand direction is no longer the dominant variable of the movement set. All movement variables are present in whatever proportion is normal in spontaneous movement. We tested whether the neuron was directionally tuned over this full movement set. Figure 8-1E shows the result of the regression fit. The fit was highly significant ($p=0.00008$). The R^2 value, however, was 0.05. Direction tuning accounted for only 5% of the total variance in this neuron's behavior. Figure 8-1F shows the R^2 values for all cells tested. The mean R^2 value was 0.08, and 63% of the cells showed a significant fit to the cosine model of directional tuning.

These results do not refute direction tuning. The majority of neurons were significantly tuned to hand direction. The results indicate, however, that when all variables are in play, a tuning to hand direction explains relatively little (about 8%) of the behavior of motor cortex neurons. Other parameters must account for the remaining variance. As described in Aflalo and Graziano (2007), a range of alternative direction tuning models returned a similar result. Whether the preferred direction was allowed to rotate depending on initial arm position, whether the baseline firing rate was allowed to vary linearly with hand position, whether the gain of the tuning curve was allowed to vary linearly with hand position, whether hand speed was included in a variety of velocity tuning models, direction tuning could account for only a small amount of the neuronal variance once a full set of movements, covering the full workspace of the arm, was taken into consideration.

As an analogy, consider a leaf in a turbulent river. What factors account for its behavior? To simplify the problem a scientist might take the leaf out of the water and put it in a vacuum chamber. All sources of variance have been minimized except gravity, and therefore the behavior of the leaf is attributable to gravity with an R^2 value approaching 1.0. This experiment correctly identifies gravity as a contributing factor, but it does not provide a useful model of the real-world behavior of the leaf in the river. In that condition, gravity contributes to a small percent of the variance while other factors drive the remaining variance. In the same manner, for neurons in motor cortex, if all sources of variance are minimized except hand direction then the neurons will appear to be tuned primarily to hand direction. During normal movement, when all variables are in play, hand direction accounts for only a small proportion of the total variance. Thus, although the neurons are tuned to hand direction, this tuning does not provide a primary or first-order description of the activity of the neurons during normal behavior.

TUNING IN 8-D POSTURE SPACE

Electrical stimulation of motor cortex on a behaviorally relevant time scale evokes movements of the arm that terminate in specific joint configurations. For example, stimulation of one location in cortex might bring the hand to the mouth with the elbow in lower space, the forearm supinated such that the palm faces the mouth, the wrist slightly flexed, the fingers in a grip posture, and the mouth open. The effect of stimulation is not merely to bring the hand to a specific end point, but to bring the many joints of the arm to a specific final configuration or posture. We therefore hypothesized that each neuron in motor cortex should be tuned to posture in the following sense. The neuron should fire most during movements of the arm that terminate at or near a preferred posture and should fire less during movements that terminate far from that preferred posture. In this hypothesis, the firing of the neuron helps cause the arm to move to that posture.

To test this hypothesis, it is necessary to work within the highly dimensional space of arm posture. Eight degrees of freedom of the arm were monitored, including grip aperture and seven joint angles. These angles included the wrist flexion, wrist adduction, forearm pronation, elbow flexion, shoulder elevation, shoulder azimuth, and shoulder internal rotation. These eight degrees of freedom define an eight-dimensional (8-D) posture space. A specific multijoint posture of the arm corresponds to a specific point in this 8-D space. Each movement of the arm can be described as a trajectory through this 8-D space.

For each neuron studied, we tested whether the neuron fired most during movements that terminated near a specific preferred point in 8-D posture space. In this analysis, the firing rate was modeled as a Gaussian function. The peak of the Gaussian was located at the preferred point in posture space. Movements that terminated at or near that preferred point in posture space should be associated with a high neuronal firing rate, and movements that terminated at a point far from the peak of the Gaussian should be associated with a low neuronal firing rate. Despite the eight dimensions, therefore, the concept is quite simple. The Gaussian tuning is a conceptually straight-forward way to embody the hypothesis of a preferred final posture.

The equation for the Gaussian model is not necessary to understand the basic concept of the analysis described below. For those who wish to know the details, however, the formal equation for the Gaussian fit was:

$$\text{Firing rate} = A e^{\sum_{i=1:8} \frac{(x_i - p_i)^2}{2\sigma_i^2}} + B$$

The firing rate refers to the mean neuronal activity during a particular movement; x_1-x_8 refer to the eight coordinates of the end point of the movement; P_1-P_8 refer to the coordinates of the peak of the Gaussian; the standard deviations of the Gaussian around that peak are indicated by $\sigma_1-\sigma_8$; the height of the Gaussian is given by A; and the floor of the Gaussian is B.

For each neuron, we used a nonlinear regression technique to fit this Gaussian model to the data. The analysis found a specific posture of the arm that the neuron preferred. The analysis also yielded an R^2 value indicating how much of the variance in the neuronal data could be explained by the fit. To avoid inflating the R^2 value with the addition of eight fitting parameters, we used the standard adjusted R^2 metric that takes into account the number of regressors (Cohen et al., 2003; see also Aflalo and Graziano, 2007, for statistical simulations used to confirm that the additional regressors did not inflate the R^2 values).

Figure 8-2 shows data from four example neurons. Each column shows data from a different neuron. In Column 1, panel A1 provides information about the preferred posture of neuron 1. In total, 380 movements of the arm were tested for this neuron. Each stick figure in panel A1 shows the configuration of the arm at the termination of a movement. The stick figures represent only the 10% of movements that terminated closest to the preferred posture. The distance metric here is not hand position. It is the distance in the 8-D space of the arm's posture. The thirty-eight stick figures in panel A1 therefore show the postures that were most similar, in their joint configuration, to the preferred posture. Panels B1 and C1 show a side and top view of the same postures. This figure gives some sense of the preferred posture for the neuron. The preferred posture involved a raised elevation of the shoulder lifting the hand mainly into upper space, a range of azimuth angles that placed the hand in a band of frontal space, a partially extended elbow that placed the hand away from the body, an extended wrist angle, and a grip aperture (not shown) that was on average 2 cm. The posture is similar to a natural reach to grasp or manipulate a small object in upper frontal space.

More results from the same neuron are shown in Figure 8-2D1. Here rasters of neuronal activity are displayed. Each row of dots shows the firing of the neuron during a movement. Because different movements were of different durations in the naturalistic movement set, these rows in the raster display are of different lengths. The first raster display shows data from the 10% of movements that terminated nearest to the preferred posture. The second raster display shows data from the 10% of movements that terminated farthest from the preferred posture. These rasters show that the firing rate was variable from movement to movement, but that the neuron fired consistently more during movements that terminated near the preferred posture and fired consistently less during movements that terminated far from the preferred posture.

The full tuning curve of the neuron is displayed in Figure 8-2E1. The Y axis shows the firing rate of the neuron during each movement. The X axis shows the distance between the final posture of each movement and the preferred

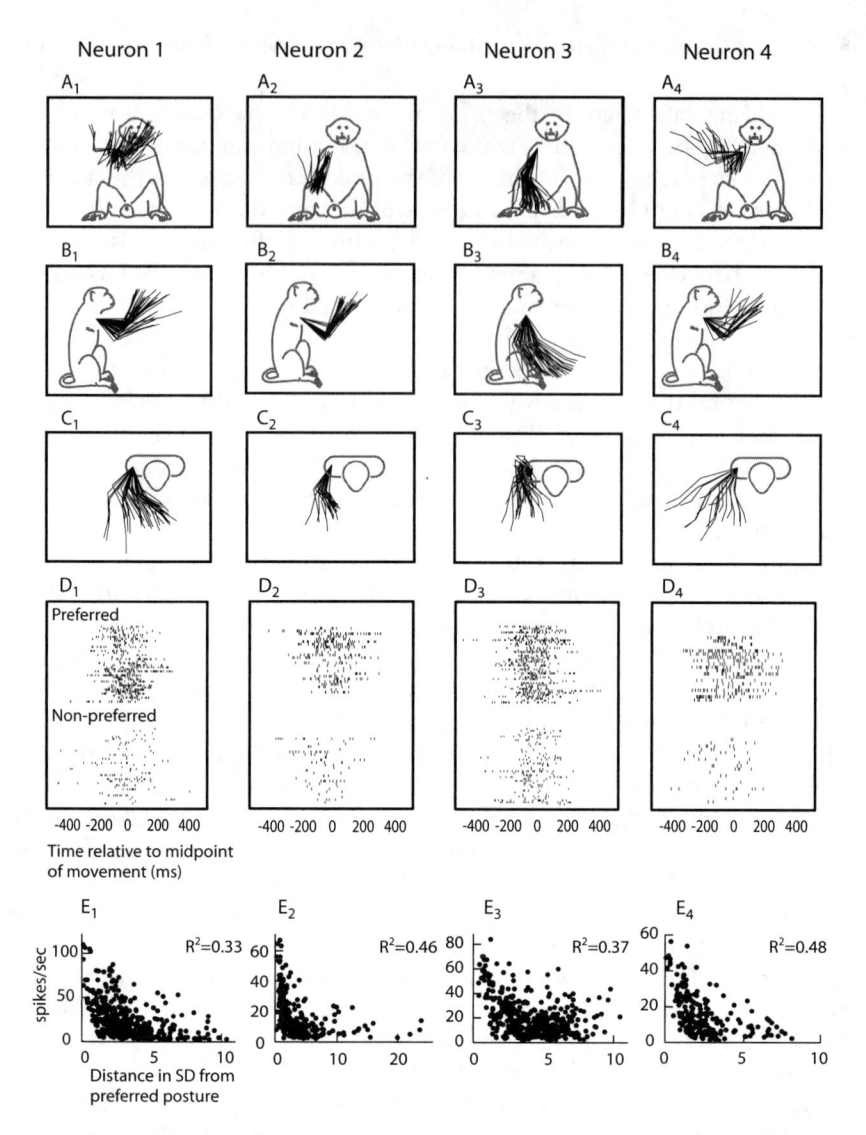

Figure 8-2: End-posture tuning of four example neurons. **A1–4.** Stick-figure drawings showing front view of the end postures for the 10% of movements that terminated nearest to the preferred posture. Three joints are shown: shoulder, elbow, and wrist. The schematic monkey drawing indicates approximate scale and orientation. **B1–4.** Same as A but side view. **C1–4.** Same as A but top view. **D1–4.** Rasters showing high neuronal activity during the 10% of movements that terminated nearest to the preferred posture, and low neuronal activity during the 10% of movements that terminated farthest from the preferred posture. **E1–4.** For each neuron, the preferred multi-joint posture was determined by regression analysis. The final posture of each movement was compared to the preferred posture. The Pythagorean distance between them was calculated in 8-dimensional posture space. Distance was measured in units of standard deviations of the Gaussian tuning function, in order to express all 8 dimensions in posture space in equivalent units. This distance is plotted on the X axis and firing rate during the movement is plotted on the Y axis. The R^2 values indicate the closeness of fit between the Gaussian model and the data for each neuron. Adapted from Aflalo and Graziano (2007).

posture of the neuron. Again, the distance metric is not hand position; it is the distance in terms of joint configuration, or 8-D posture space. A distance of 0 on the X axis indicates that the movement landed exactly on the preferred posture of the neuron. A high firing rate is therefore expected in that case. A large distance on the X axis indicates that the movement ended far from the preferred posture of the neuron. A lower firing rate is therefore expected. This is precisely the case. On average, movements that terminated near the preferred posture had higher firing rates, and movements that terminated progressively farther from the preferred posture had progressively lower firing rates. About 33% of this neuron's variance was attributable to a Gaussian tuning to the preferred posture ($R^2=0.33, p < 0.0001$).

All four example neurons in Figure 8-2 show a similar pattern. Each neuron preferred a different posture and was tuned to that posture in a roughly Gaussian fashion. In total forty-six neurons were tested in this fashion. The mean R^2 value across the sample of neurons was 0.36. The tuning to a goal posture of the arm therefore explained slightly more than one third of the variance in neuronal activity.

To understand the meaning of this result it is necessary to recall that the movement set used for this test was a global movement set that included all recorded arm movements. The set was not preselected to enhance the effect of one or another movement parameter. All parameters relevant to normal spontaneous movement were in play. The factors of hand speed, hand direction, the length of the movement, the curvature of the movement, acceleration, force applied by the arm, arm posture, and the behavioral meaning of the action all varied from movement to movement. If neurons in motor cortex controlled all of these variables equally, then each one would be expected to explain a small proportion of the total variance. Movement of the arm to a preferred posture, however, explained a disproportionately large share of the variance, capturing 36% of the total. Although undoubtedly tuned to many control variables, the neurons in motor cortex were better tuned to the goal posture of the arm than to any other variable that we tested. For a complete description of the alternative tuning models that were tested, see Aflalo and Graziano (2007).

TUNING IN POSTURE SUBSPACES

As described above, neurons in motor cortex were tuned to posture in the sense that a neuron fired most during movements that brought the arm to a preferred posture and fired progressively less during movements that brought the arm to postures progressively distant from the preferred one. The posture space in which the neuron was tuned was the 8-D space of joint angles. Yet in plotting the tuning curves for neurons, we noticed that neurons were typically robustly tuned for some joints and not for others. In the first example neuron shown in Figure 8-2, the neuron was narrowly tuned to shoulder elevation and to elbow flexion but broadly tuned to shoulder azimuth. These tuning properties can be seen in the dispersion of arm postures shown in 8-2A, B, and C. There is little dispersion in shoulder elevation or elbow flexion, and a wide dispersion in shoulder azimuth. In this sense it appeared that the neuron was not tuned to a complete 8-D arm posture but instead was tuned to a subset of the joints.

A more formal method of examining this unequal tuning to different joints is to calculate partial R^2 values. The partial R^2 value for a particular joint is the amount of R^2 that is lost by dropping that joint from the regression analysis. If a neuron is equally tuned to each of the eight joint dimensions, then the eight partial values should be approximately equal. Most neurons, however, did not have equal partial values for all eight joints. Instead neurons tended to be better tuned to some joints than to others. On average, neurons reached statistical significance in their tuning to four of the eight joints.

Different neurons were tuned to different subsets of joints. For example, one neuron might be tuned to all three shoulder angles, forearm pronation, and grip aperture. Another neuron might be tuned to elbow flexion and wrist flexion. Another neuron might be tuned only to the most distal joints, the wrist and grip. Another neuron might be tuned to a mixture of distal and proximal joints. Every neuron was significantly tuned to at least two of the eight joints; no neuron was significantly tuned to all eight joints. It was as if each neuron, in firing, was specifying some part of an arm posture, instructing several joints to move to a particular set of angles; and only many neurons together could specify a complete arm posture.

MATCH BETWEEN POSTURES PREFERRED BY NEURONS AND POSTURES EVOKED BY ELECTRICAL STIMULATION

Immediately after recording from neurons at a cortical site, we electrically stimulated the same cortical site through the same electrode. This procedure allowed us to compare the tuning of each neuron to the movement evoked by electrical stimulation. Figure 8-3 shows data from an example neuron. First the preferred posture of the neuron was found using the 8-D regression analysis described above. In Figure 8-3A, each gray stick figure shows the configuration of the arm at the termination of a movement. Displayed are the 10% of movements that terminated closest to the neuron's preferred posture, providing a sense of the range of arm configurations generally preferred by the neuron.

After recording from this neuron, at the same cortical site, we electrically stimulated for 500 ms and obtained a stimulation-evoked posture. Twenty stimulation trials were tested. The thick black stick figure in Figure 8-3A shows the mean configuration of the arm at the termination of the stimulation train. In this case, the stimulation-evoked posture was within the range of postures preferred by the neuron. The match seems close but not exact. It is quite difficult, however, to assess the closeness of the match intuitively, given the complexities of an 8-D posture space. Is the match displayed in Figure 8-3 much better than expected by chance, or is it close to chance occurrence? A more

8. Natural Neuronal Properties and Stimulation-Evoked Movement

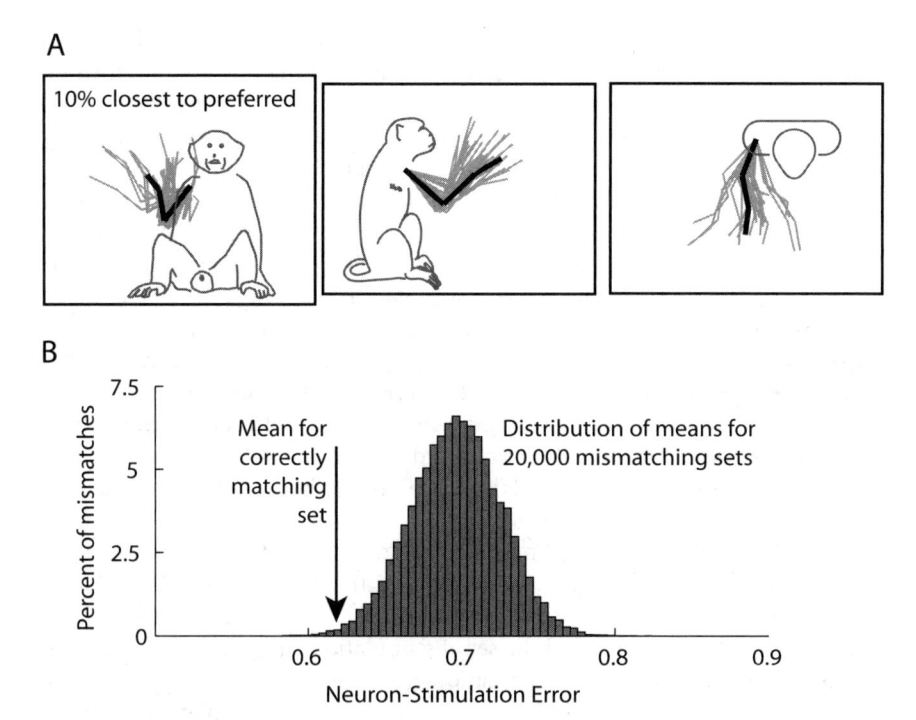

Figure 8-3: Match between tuning properties of single neurons and stimulation-evoked movements. **A.** Results from one example neuron. The gray stick figures show the final arm postures for the 10% of movements that terminated closest to the neuron's preferred posture. The black stick figure shows the mean arm posture evoked by electrical stimulation of the same cortical site. **B.** Statistical test for the match between neuronal tuning and stimulation-evoked posture. For each neuron-stimulation pair, an error was calculated based on the distance in joint space between the neuron's preferred posture and the stimulation-evoked posture. The mean for the neuron-stimulation pairs was 0.62. The analysis was then repeated with neurons and stimulation sites randomly mismatched. A total of 20,000 mismatched sets were tested. These randomly mismatches sets on average had a larger Neuron-Stimulation Error than the correctly matched set. Based on data from Aflalo and Graziano (2007).

quantitative approach is necessary to understand the remarkable closeness of this match. We used a "bootstrap" analysis, as follows.

First, for each neuron, we selected the four joints to which it was most tuned (the joints for which the posture tuning had the largest partial R^2 values). For example, if a neuron was highly tuned to movements that terminated at a specific elbow angle, then by hypothesis stimulation of the same cortical site should evoke movement to that elbow angle. But if the neuron was not well tuned to elbow angle, then the neuron should not be expected to closely match the stimulation-evoked elbow angle.

Using the space of the four best joints for the neuron, we then calculated a distance between the neuron's preferred posture and the posture obtained on stimulation. The distance was a Pythagorean or straight-line distance in 4-D space, in which each joint dimension was normalized to a scale of 0 to 1. For each neuron, therefore, a single number was obtained, the neuron-stimulation error. The smallest possible error was 0, indicating a perfect match between the neuron's tuning and the stimulation result. The largest possible error was 2, indicating as great a mismatch as possible in four joint dimensions. For the example in Figure 8-3A, the neuron-stimulation error was 0.49. The mean neuron-stimulation error among all neurons tested was 0.62.

We then randomly mismatched the neurons and the stimulation sites and recalculated the mean neuron-stimulation error. In the null hypothesis, if the properties of neurons have no specific match to the effects of stimulation, then this random mismatch of neurons and stimulation sites should have little impact on the result. The neuron-stimulation error should remain unchanged. However, if a specific match does exist between neuronal properties and the effect of stimulation, then a random mismatch of neurons and stimulation sites should remove this correspondence, and the mean neuron-stimulation error should increase. We tested twenty thousand random mismatches between the set of neurons and the set of stimulation sites. For each mismatch we calculated a mean neuron-stimulation error. Figure 8-3B shows the resulting distribution. The arrow indicates the mean neuron-stimulation error for the correct pairing of neurons and stimulation sites. The distribution obtained by random mismatches is 99.6% above this level. The match between neuronal properties and stimulation effects is therefore highly unlikely to be a result of chance because a chance pairing of neurons and stimulation sites results in a worse match in almost every instance. This result indicates a significant correspondence between neuron properties and stimulation properties at the same cortical site.

INTERPRETATION: NEURONS IN MOTOR CORTEX CAUSE USEFUL MOVEMENTS

In summary, neurons in motor cortex are significantly tuned to posture, in the sense that a neuron is most active during movements that terminate at or near a preferred posture. This tuning to posture emerges only when tested over a broad range of arm postures. It explains approximately one third of the variance in neuronal activity. Neurons however are not tuned to an entire arm posture but instead tend to be tuned to partial postures, on average showing good tuning to four arm joints, with different neurons tuned to different subsets of joints. When the posture preferred by a neuron is compared to the posture evoked by electrical stimulation of the same cortical site, the two postures tend to match. Because the match is in a highly dimensional joint space, the closeness of the match is difficult to assess intuitively. However, statistical methods suggest that across the population of neurons and stimulation sites, the match is close with a high degree of statistical significance.

8. Natural Neuronal Properties and Stimulation-Evoked Movement

Our interpretation of these results is that when a neuron in motor cortex becomes active, it influences the network to which it is connected in such a way as to bias the system toward producing a specific movement. This movement is a complex fragment learned by the network from the statistics of normal behavior. The movement represented by neurons at a site in cortex can be uncovered either by electrical stimulation or by plotting tuning curves during single-neuron recording. Both techniques result in somewhat approximate answers. The stimulation technique presumably blurs the properties of many nearby neurons and produces abnormal temporal dynamics. The correlative single-neuron technique is much more fraught with interpretational difficulties and is unstable, providing radically different results depending on the specific movement set used to test the neurons. With a diverse movement set that is not preselected to enhance one or another type of movement parameter, it is possible to obtain a tuning curve from a neuron that approaches the result of electrical stimulation at the same cortical site.

If neurons in motor cortex are wired to cause fragments of normal behavior, why then are the neurons more tuned to the final posture of the arm than to other variables, and why does electrical stimulation evoke movements of which the final posture of the arm is a salient characteristic? Normal behavior of the arm tends to involve underlying postures. For example, interaction between the hand and mouth involves a basic underlying joint configuration. The upper arm is vertically downward, the elbow is bent, the forearm is rotated to orient the palm toward the face, the wrist is straight or partially flexed, and the grip is typically closed. In the richness and complexity of interaction between the hand and mouth, this underlying posture is conserved within narrow limits. Other common actions in the repertoire, such as outward reaching, manipulation of items in central space, and defensive actions, to a greater or lesser extent also utilize convenient underlying postures. In the present interpretation, because acquiring and maintaining useful, canonical postures of the arm is a prominent part of normal monkey behavior, and because neurons in motor cortex cause common fragments of normal behavior, therefore the neurons cause actions for which posture plays a large role.

It has been suggested that movement might be accomplished by means of a postural control mechanism (e.g., Feldman and Latash, 2005; Rosenbaum et al., 1995; Shadmehr, 1993). In that hypothesis, the control mechanism specifies a series of postures. The arm, moving from one posture to the next, traces out the desired action. Our results on electrical stimulation and single-neuron recording do not argue for that hypothesis. We find that posture is only one control variable represented by neurons in motor cortex and specified by stimulation of motor cortex. In our hypothesis, if posture is represented prominently in motor cortex, it is only because of its prominent use in many common behaviors. It is important to distinguish between postural control as a general method of guiding movement (a controversial idea that may or may not be correct) and the control of certain distinctive postures that are highly useful in a monkey's normal behavioral repertoire (a basic fact of the repertoire that can easily be observed).

Chapter 9

The Movement Repertoire of Monkeys

INTRODUCTION

Much work on motor control in primates over the past thirty years has focused on the control of reaching. This area of research is vast, including physiological work in monkeys, psychophysical work in humans, and computational modeling (e.g., Georgopoulos et al., 1982; Li et al., 2005; Rosenbaum et al., 1995; Shadmehr and Moussavi, 2000). Reaching, in the experimental sense, is usually a cursor-like transport of the hand from one location to another. Much work has also focused on grasping and especially the preshaping of the hand in anticipation of grasp (e.g., Jeannerod, 1986; Murata et al., 1997). Before the start of our electrical stimulation experiments, I shared the same perspective as the rest of the field. In studying the properties of neurons in the motor and premotor cortex, I believed that I was studying the mechanism that controls reaching. Once my colleagues and I began our stimulation experiments, we sat day by day in front of a monkey and watched it perform thousands of stimulation-evoked movements that came straight from the normal monkey repertoire. Few of these movements bore any relation to reaching or grasping in the usual experimentally studied sense.

The stimulation-evoked movements forced us to become aware of a disconnect in the literature. The monkey motor system has been intensively studied for 130 years, since Ferrier's experiments in 1874 on the macaque motor cortex. Yet little or nothing is known about the overall organization of the movement repertoire of monkeys.

One summer morning in 2003, my graduate student Dylan Cooke and I set out to study the movements of monkeys. We packed a video camera and tripod in a duffle bag and took a train to the Bronx Zoo, which generously allows its patrons to film the exhibits. We filmed a range of primates including silvered leaf monkeys (*Presbytis cristata*), squirrel monkeys (*Saimiri sciureus*), gibbons (*Hylobates lar*), gorillas (*Gorilla gorilla gorilla*), gelada baboons (*Theropithecus galada*), and for diversity of order some squirrels (*Sciurus carolinensis*) that were freeloading on the zoo garbage cans. The zoo animals were housed in seminatural group enclosures. We were able to film complex behavior including climbing, playing, grooming, foraging, fighting, and so on. Much of this video footage was analyzed frame by frame in an attempt to construct a general, qualitative description of the normal movement repertoire of monkeys.

Several years later an intrepid member of the lab, Nico Macfarlaine, spent a summer on Cayo Santiago, an island populated by semiwild rhesus monkeys, and collected a large and extremely informative film library.

This chapter is intentionally descriptive. It does not present experiments. It describes some of the generalizations and insights that resulted from the video analysis. Any field of study often begins with a general description that is then subjected to increasingly specific hypothesis testing. The action repertoire of monkeys lacks this most basic level of description, and therefore this chapter intentionally steps back to take a broad perspective on the issue of action repertoire.

Perhaps the most striking feature of the movement repertoire of monkeys, or of any animal that we observed, was its breakdown into action modes and submodes between which the animal frequently switched with minimal overlap. It was as if the animal had a dial inside of it that pseudorandomly switched from one setting to another. At least in the case of monkeys, a surprisingly straightforward correspondence may exist between action modes and areas of the cortex in which those modes are emphasized. Once one understands the natural divisions in the animal's behavioral repertoire, the natural divisions in the motor system become more interpretable.

ACTION MODES OF MONKEYS

The principal action modes that we observed in monkey behavior are hardly a surprise. They are common in our own behavior and therefore seem intuitively obvious. They are listed in Figure 9-1. This list is not complete. It leaves

```
Acting on objects (45%)  ⇐ Reaching (3%)
                            Manipulating (20%)
                            Hand-and-mouth interaction (22%)
Locomotion (3%)
Exploratory gaze (52%)
```

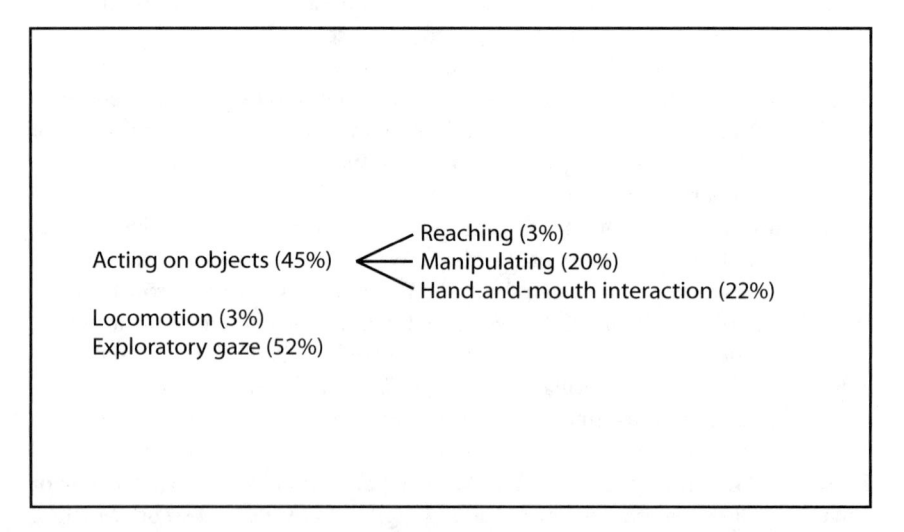

Figure 9-1 Percentage of waking time monkeys were observed to spend within five main action modes. This list is far from complete but provides a rough summary of much of monkey behavior.

9. The Movement Repertoire of Monkeys

out many behavior types but captures the classes of behavior that occupy the majority of a monkey's time. In the respects described here, monkey behavior was similar across the different species filmed including silvered leaf monkeys, fascicularis macaques, rhesus macaques, and squirrel monkeys. Most of the analysis was performed on the macaques housed in our laboratory because that analysis provided the most direct comparison to the results of our electrical stimulation experiments. The numbers given below are from that group.

Acting on objects occurred almost always when the animal was seated on its haunches and the arms and hands were free. This mode contained three main submodes: reaching to an object, manipulating an object in central space with one or both hands, and hand-and-mouth interaction in which an object was held at the mouth and manipulated, or chewed, or taken out of the mouth for inspection.

Locomotion involved far more than a simple cyclic stepping of the limbs. It typically involved a complex, constantly changing placement of the hands and feet on opportune surfaces or objects.

Exploratory gaze involved moments of examining the larger environment beyond the immediate objects with which the monkey was interacting. Changes in gaze were of course not limited to this particular action mode. All action modes involved changes of gaze. During reaching to an object, for example, the monkey's gaze shifted transiently to the object. During locomotion, the gaze seemed to move to specific goal locations. Exploratory gaze was distinct in that it appeared to involve an exploration of the distant environment and a cessation of other actions.

Typically an animal switched rapidly among these different action modes. For example, a monkey might sit on its haunches and manipulate a piece of food with both hands in central space; then lift the piece of food to the mouth and tear at a corner of it with its teeth; then lower the piece of food to central space and manipulate it between the hands again; then lift it to the mouth for further hand-and-mouth manipulation; then pull the object away from the mouth again for bimanual manipulation; then stop all manipulatory action, hold the food still, lift its head, and begin a sequence of exploratory gaze, looking first one way and then another; then lower its gaze to the object and return to manipulatory behavior; then drop the food and begin to locomote, walking or running or climbing; then stand still on all fours and begin another exploratory gaze sequence; then sit, reach out to a nearby play object such as a chip of wood, grasp and bring the object to the space just in front of the chest and begin to manipulate it. Figure 9-2 shows a typical time sequence in which a monkey switched from one type of behavior to another.

On average monkeys spent about 45% of their time in object interaction mode (3% reaching to an object, 20% manipulating, 22% hand-and-mouth interaction), 52% in exploratory gaze mode, and 3% in locomotion mode. Probably the low proportion of time in locomotion reflects the condition of captivity. The episodes of each action mode were brief. Periods of object interaction mode lasted on average 6.9 +/− 6.1 sec; exploratory gaze lasted an average

▨▨▨▨ Reaching

☐ Manipulating

■ Hand-and-mouth interaction

▧▧▧▧ Locomotion

■ Exploratory gaze

Typical sequence

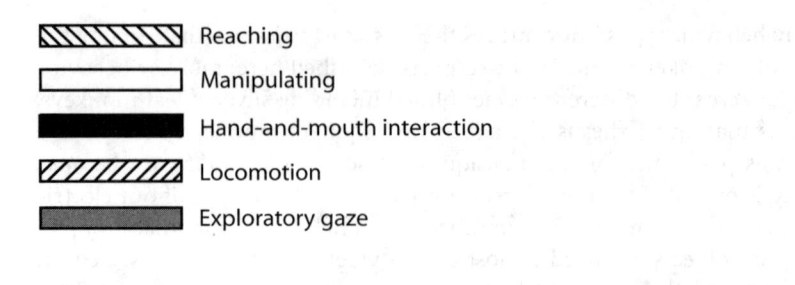

|———|

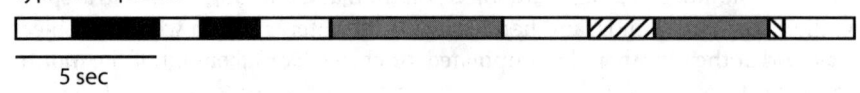

5 sec

Figure 9-2 Monkeys appeared to switch incessantly between several dominant action modes. A typical sequence of action modes for a monkey is shown here.

of 5.1+/−7.2 sec; and climbing and walking excursions had a mean duration of 2.3+/−7.0 sec. The impression was of a constant changing from one mode to the next.

Some overlap did occur between action modes. For example, exploratory gaze did sometimes occur while the animal continued to manipulate an object in the hands. More commonly, however, exploratory gaze was accompanied by a temporary cessation of other ongoing actions. One speculation is that attention places a limit on the number of tasks the animal can perform simultaneously. In this speculation, an animal may be continually aware of its larger surroundings, but during brief periods the animal pays more attention to the task of gathering information about its surroundings through head and eye movements and during these times, because of a limited attentional resource, it stops its other ongoing actions. In this view, the animal can perform more than one action at a time if the actions are relatively undemanding, but it tends to perform only one complex or attentionally demanding action at a time, and therefore the behavior of the animal has the appearance of switching rapidly from mode to mode.

The principal action modes may partially map onto the main sectors of cortex that control movement. As detailed in Chapter 7, stimulation in the lateral motor cortex, including the primary motor and caudal premotor areas, evokes actions that most resemble the interaction with objects. Stimulation in the medial motor cortex, including SMA, evokes actions that most resemble the complex, multilimb movements and body adjustments during locomotion. Stimulation in the FEF evokes gaze shifts (Bruce et al., 1985; Chen, 2006). These areas of cortex presumably are not exclusive modules; they share function and interact extensively. Yet it is possible that the cortical motor system in primates has organized broadly around the general statistical structure in the behavioral repertoire. This issue of the mapping of behavioral repertoire onto the cortical sheet is addressed in greater detail and with a formal quantitative model in Chapter 10.

GRIP

More than anyone, Napier (1956) contributed to the study of grips and their evolution. Most work on grip emphasizes two common categories of primate grip: a precision grip and a power grip. The precision grip, between the forefinger and thumb, is considered to be a more recent evolutionary product and is fully expressed only in some primate species (Marzke and Marzke, 2000). Yet grip is much more diverse than these two examples, in human and monkey behavior as well as in the behavior of other animals. This section offers a different perspective on grip that emerged from our observations and that in no way conflicts with the more standard description of primate grip.

One essential feature of grip that becomes apparent on examining animal behavior is that a single grip on an object is not sufficient for most useful purposes. Monkeys and humans sometimes grip objects for the purpose of examining them, in which case a single grip is sufficient while the hand rotates the object for visual inspection. But examining an object is a limited advantage. One wants to act on the object, and most grip is used in this context. Ideally one needs two grips, at two different locations on the object, such that a differential force can be applied that will alter the object. The interesting story of grip is therefore mainly about how to apply two grips to the same object.

Arguably the simplest action on an object is biting it and tearing off a piece. The mouth can be said to be the original gripper. Consider a predator such as a dog eating a carcass. If the carcass is large enough then gravity does the job of stabilizing the carcass while the dog grips and tears with its mouth. A small carcass, however, is not easily manageable in this manner, and the dog resorts to a double grip. One grip is between a forepaw and the ground, pinning and stabilizing the object. The other grip is between the teeth. A predator such as a dog or lion therefore can be said to have a two-grip manipulation of objects that involves an interaction between the forepaw and the mouth.

Animals that sit on their hindquarters can achieve a more controlled grip by opposing the two forepaws. Squirrels for example sit on their haunches and hold a seed (or French fry) between the pads of their forepaws while chewing it. Here again is a two-grip action on an object. The differential force introduced between the mouth grip and the two-paw grip allows pieces of the food to be bitten off. Furthermore, the object can be rotated or adjusted in the two-paw grip to aim different parts of it at the mouth. Manipulation of objects in this framework is still focused on the mouth and interactions between the forepaws and the mouth.

In primates grip becomes more complex yet. Opposability is possible not only between the two forepaws, but also between separate elements on a single forepaw. This advance is major because it allows a two-grip manipulation that does not include the mouth. Each hand can independently grip an object. A differential force can then be applied between the two hands, twisting the object, tearing it, or breaking it. The manipulation of objects can become dissociated from the mouth.

Primates of course retain the simpler versions of grip. In our videos we saw examples of monkeys biting pieces of objects directly from branches or food racks, effectively in a single-grip manipulation; biting pieces of an object pinned to the ground beneath one forepaw, like a dog; holding an object between the two palms while biting it, like a squirrel; holding an object in one hand while biting it; and twisting or tearing an object between the two hands.

Despite the somewhat hand-like appearance of the monkey foot, we never saw a monkey manipulate an object with the foot, in the sense that the foot never participated in a two-grip application of sheer force to an object. Instead the foot was sometimes used as a storage device. When a monkey was sitting and manipulating an object with its hands, it sometimes placed the object in the foot and then reached for and manipulated a second object in the hands. Once the second object was dropped or consumed, the monkey then took the first object out of its foot for further manipulation. Sometimes a monkey walked with a stored food item still held in the curled foot. In this sense the foot is capable of grip but showed no evidence of manipulation.

In humans, grip becomes immensely more complex. It is no longer limited to specific anatomical grippers. Any two body parts that can be physically opposed will be used opportunistically for grip. Consider a cigarette grip between the third and fourth finger; a newspaper grip between the upper arm and torso; a soda bottle grip between the knees when sitting; or an accidentally-falling-book grip between the open palm and the chest. It is as if the general concept of opposability of body parts has been mastered by the motor system.

One remarkable increase in complexity in human grip is the ability to achieve a two-grip manipulation of an object within one hand. For example, consider the one-handed-bottle-opening behavior in which the neck of the bottle is gripped between the palm and digits 3–5, the cap of the bottle is gripped between forefinger and thumb, and a differential force is applied between the two grips to unscrew the lid.

Electrical stimulation of the monkey motor cortex evoked complex grips in two cortical zones: the hand-to-mouth zone and the manipulation zone. The range of grips evoked by stimulation of these cortical zones resembled the range of grips observed in monkey behavior. For example, stimulation-evoked grips included a bringing together of the pad of the thumb with the pad of the forefinger in a precision grip; a bringing together of the pad of the thumb with the side of the forefinger, in a variant of the precision grip common in monkey behavior; a fist-like closure of the hand in an apparent power grip; and a partial closure of the hand as though to grip a larger object. Stimulation of the manipulation zone also evoked rotations of the wrist and forearm consistent with manipulation of objects, and often a closure of the grip and a fast lateral movement of the hand that we interpreted as similar to a tearing action. Stimulation in the manipulation zone did not evoke bilateral movements. It did not evoke a full pantomime of a two-handed manipulation. Instead it evoked what appeared to be fragments of manipulation in one hand. The stimulation results therefore are consistent with the hypothesis that during bimanual manipulation of an object, the two hands are controlled separately by unilateral representations in cortex. It may be that this separate control allows the hands to perform different actions on the object, such as holding the object with one hand while twisting it with the other hand.

REACH AND INWARD SCOOP

During normal behavior, reach and grasp belong to a larger context of two common processes. During one process, the arm projects outward, the forearm pronates thus orienting the palm outward, the wrist extends, and the grip opens. During the second process the arm pulls inward, the forearm supinates thus orienting the palm inward, the wrist flexes, and the grip closes. The first process corresponds to reaching out toward an object to be acquired, and the second process corresponds to grasping and scooping the object in toward a more useful workspace near the body. The two processes usually overlap in time; the closing of the grip begins well before the hand has reached the object (Jeannerod, 1986).

In our stimulation studies, when we stimulated in the cortical zone that we termed the "reach-to-grasp" zone, we tended to evoke the first process of extension of the arm, pronation of the forearm, extension of the wrist, and opening of the grip. In this sense the term *reach-to-grasp* is misleading because there was no grasp at the end of the reach. Rather, we seemed to evoke a reach in preparation for a grasp, with the grip open. In contrast, when we stimulated in the cortical areas that we termed the "manipulation" zone and the "hand-to-mouth" zone, we tended to evoke the second process of retraction of the arm toward the body, supination of the forearm, flexion of the wrist, and closure of the grip.

LOCOMOTION

Locomotion was not a uniform type of behavior. It was an aggregate of complex, sensory-motor actions. It involved a constantly changing placement of the hands and feet on opportune surfaces in precise sequences. Perhaps because of the cluttered and vertical environments typical of a zoo enclosure and also in our primate facility, locomotion was almost never a simple matter of cyclic leg movements on a flat surface. The movements involved the simultaneous coordination of two, three, or all four limbs, the posture of the trunk and head, direction of gaze, and movement of the tail presumably for balance. Typically the head and eyes began to turn first toward a new location. The upper torso then began to turn, the arms reaching to a new set of holds. The lower body then began to turn. At the new location the hands tended to land first, then the feet. Sometimes only the upper body changed its posture, as the monkey shifted to a new set of hand holds and gained a new vantage point to examine the larger environment. The movements were therefore not always equally integrated over the entire body but often emphasized the upper or the lower body.

The hand actions used during locomotion were simpler than those used during grasp and manipulation of objects. The four fingers were usually together and either straight such as during walking on a flat surface, or slightly curved such as during walking on or leaning momentarily on a branch. During climbing, the four fingers sometimes curved into a hook that was used to hang from a narrow branch or bar. Thus the four fingers acted as a unit with varying degrees of curvature. The thumb was only occasionally used in opposition to the fingers. Only when climbing a vertical, narrow branch did monkeys use the opposability of the thumb extensively, gripping the branch between the fingers curled as a unit to one side and the thumb curled to the other side. In these respects, locomotion was different from the grasp and manipulation of objects, which involved the complex and sometimes independent use of the fingers and frequent opposition of the thumb to the fingers to form a variety of grips.

The movements that we evoked by electrical stimulation of sites in SMA resembled the set of movements typical of complex locomotion. The evoked movements were often bilateral, sometimes involved all four limbs, and sometimes included the tail curling to one side as in normal maintenance of balance. The evoked movements often included a reaching of the hands as if to lateral or upper hand holds. In these cases the fingers shaped in a simple fashion, the four fingers together in a gently curved hook. Complex grips were not typically found on stimulation of SMA. Sometimes, especially in the more posterior part of SMA, the evoked movement involved a turning of the hips to one side and a lateral reaching movement of one foot, as though the animal were shifting its stance. These evoked movements were therefore consistent with the hypothesis that one function of the SMA is to contribute to complex locomotor behavior, such as leaping, climbing, or walking through a complex environment.

DEFENSIVE ACTIONS

One of the first scientists to study defensive behavior in a natural context was Hediger (1955) who described the phenomenon of a flight zone. Hediger noted that an animal did not simply flee at the sight of a predator. Instead, when the predator had approached within a specific distance, the threatened animal moved away to reinstate the margin of safety. One of the most useful insights in Hediger's work was the realization that defense is not simply a reflexive reaction to a stimulus. Instead it is a constant process of spatial

9. The Movement Repertoire of Monkeys

computation and movement adjustment, shaping ongoing behavior in such a way as to preserve a margin of safety. Only when an extreme threat penetrates the margin of safety does the defensive mechanism trigger an overt or extreme withdrawal.

This phenomenon of a margin of safety was obvious in our video footage of monkeys. Smaller monkeys maintained a spatial separation from larger monkeys. In walking or climbing, monkeys constantly adjusted their ongoing movements to avoid potential collision with branches, bars, or other monkeys. Only in rare moments did an overt flinch occur. For example, when a baby animal climbed on an adult, the baby might put a hand in the adult's eye, evoking a sudden strong defensive reaction: a squint, blink, turning aside of the head, shrugging of the shoulder, and lifting of the hand to knock away the colliding object.

As described in Chapter 7, electrical stimulation within a specific region of the monkey motor cortex, PZ, evokes movements that closely resemble defensive reactions. Neurons in PZ respond to tactile, visual, and auditory stimuli. The sensory receptive fields of the neurons resemble bubbles of space anchored to the skin. The receptive fields are most common on and around the upper body, especially the face. As shown schematically in Figure 9-3, the receptive

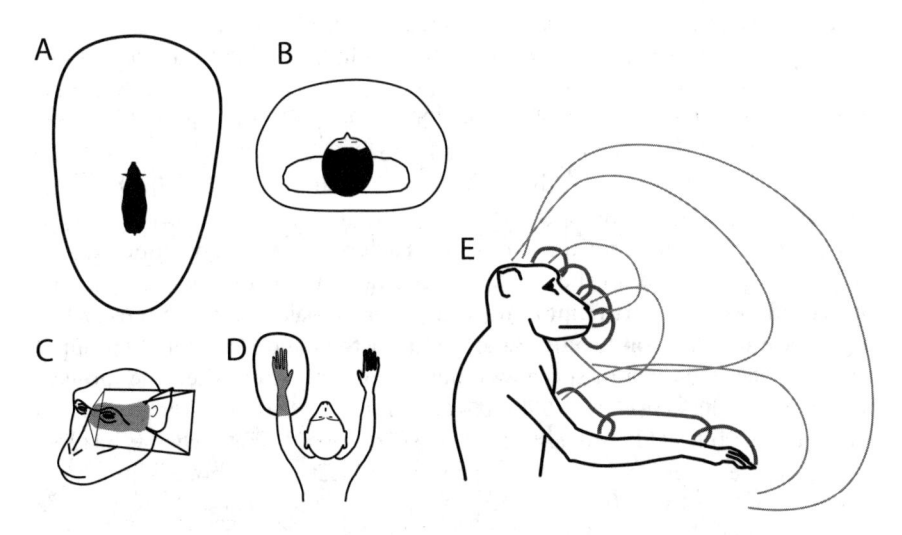

Figure 9-3 Peripersonal space. **A.** The flight zone of an animal. When a threat enters the flight zone, the animal moves away. Adapted from Smith (1998). **B.** The personal space of a human. When another person enters the personal space, the subject moves away. **C & D.** Some tactile receptive fields (shaded) and visual receptive fields (boxed) of neurons in monkey cortical area PZ. **E.** Schematic diagram of visual receptive fields in PZ. Space near the body is represented by relatively more receptive fields, and space at increasing distances from the body is represented by fewer receptive fields.

fields vary in the distance to which they extend from the body. Some extend only a few cm, whereas others can extend meters from the body. In this manner, the overlapping receptive fields represent the space around the body with relatively greater but not exclusive representation of nearby space. These spatial receptive fields around the body, and the defensive and retracting functions generated by the activity of these same neurons, provides a possible mechanism to explain flight zone and personal space.

In the current hypothesis, the defensive mechanism operates in a graded fashion. Under most circumstances, neurons in PZ continuously respond to sensory stimuli near the body and subtly bias ongoing action away from the objects in personal space. A more salient stimulus such as a looming object may evoke a greater neuronal response and an overt withdrawal. An extreme threatening stimulus may evoke intense activity in PZ and result in an extreme protective reaction. The lower end of this functional range would manifest itself as a subtle behavioral avoidance such as is associated with maintaining a margin of safety. The upper end of this functional range would manifest itself as flinching.

COMPARING THE REPERTOIRE OF POSTURES IN SPONTANEOUS BEHAVIOR TO THE POSTURES EVOKED BY STIMULATION

To explore further the possible match between the movement repertoire of monkeys and the movements evoked by stimulation of the monkey motor cortex, we compared the statistical distribution of hand locations in both sets of movements (Graziano, Cooke, et al., 2004). As shown in Figure 9-4A, the space in front of the monkey was divided into nine imaginary zones, each one 10 cm across. In Figure 9-4B, the diameter of the circles indicates the percentage of time that the hand spent in each zone during spontaneous behavior. The pattern was similar for the two hands. Each hand spent most time in location 5, directly in front of the chest. This central space was used to manipulate objects and as a support point against the floor or walls while climbing, walking, or leaning. A second common location for the hand was zone 2, in upper central space. This area of space was most commonly used when the monkey held an object up at eye level to investigate it more closely, or held the object to its mouth to bite it. It was also used when the hand scratched the head or pushed at the cheek pouches. A third common location was zone 8 and 9, the lower space directly in front of and lateral to the body. These areas of space were used mainly to support the body's weight, such as when the monkey leaned to the side while sitting or climbing.

Figure 9-5 compares the hand positions evoked by stimulation and the hand positions observed during spontaneous behavior. The black bars in the graph show the percentage of stimulation sites that caused the hand to move into each spatial zone. Zone 5, just in front of the chest, and zone 2, near the mouth, were particularly well represented. The gray bars show the proportion

9. The Movement Repertoire of Monkeys

A Definition of Spatial Zones

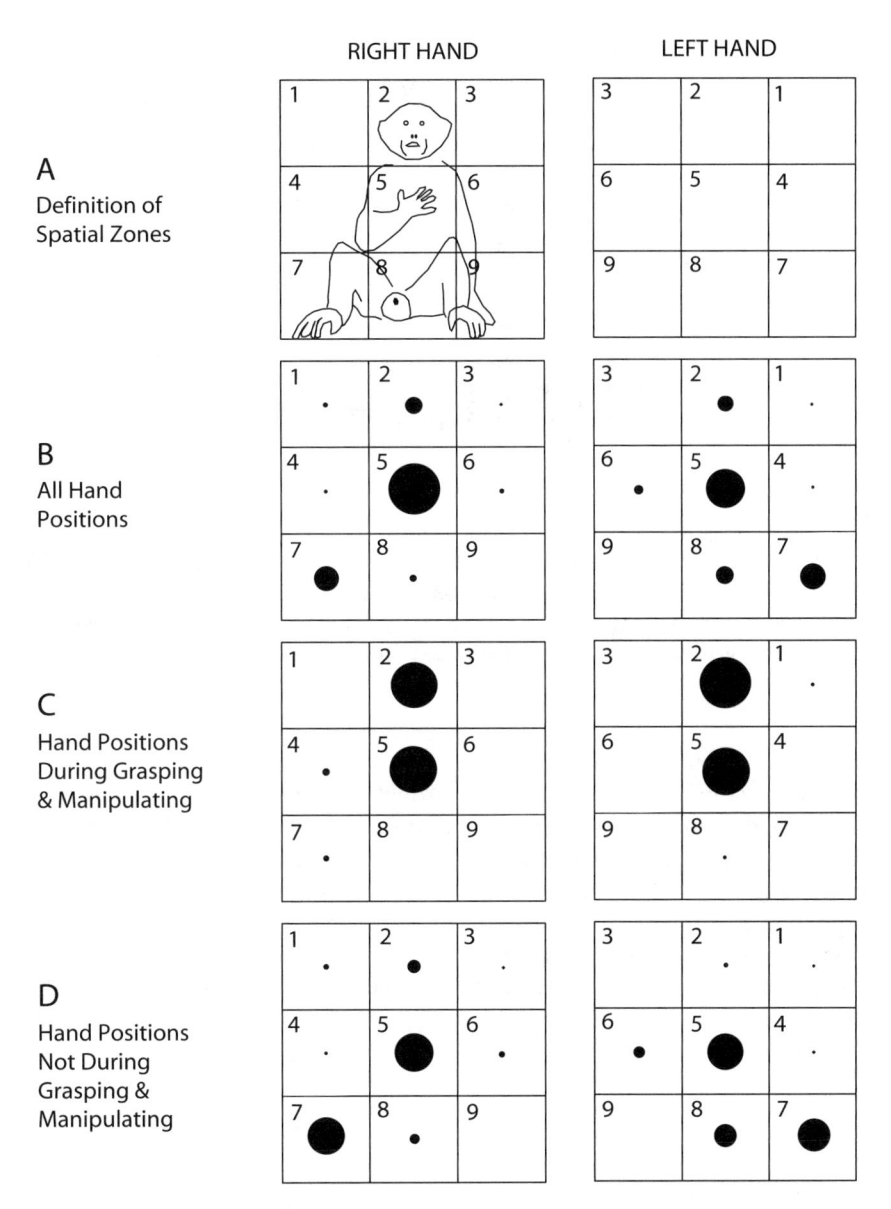

B All Hand Positions

C Hand Positions During Grasping & Manipulating

D Hand Positions Not During Grasping & Manipulating

Figure 9-4 Distribution of hand positions in the spontaneous behavior of one laboratory monkey in its home cage. Diameter of circles is proportional to the amount of time that the hand spent in each zone. Adapted from Graziano, Cooke, et al. (2004).

of times the hand moved into each zone during the monkey's spontaneous behavior in the home cage. The spontaneous behavior closely matched the stimulation-evoked behavior in statistical distribution. Those hand positions that were common in the monkey's spontaneous repertoire were also commonly evoked by stimulation of motor cortex.

[Graph showing frequency percentages across cortical zones 1-9, with black bars representing % of stimulation sites and white bars representing % of times hand entered each spatial zone during spontaneous behavior]

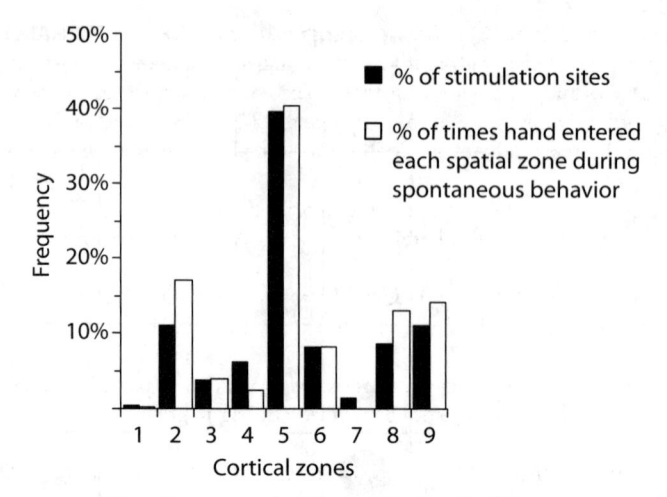

Figure 9-5 Comparison of spontaneous behavior and stimulation-evoked behavior. The height of each black bar indicates the percentage of cortical sites (of 270 total sites) for which electrical stimulation caused the hand to move into each spatial zone. See Figure 9-4 for definition of spatial zones. The white bars show the proportion of times that the hand entered each spatial zone during spontaneous behavior in the home cage. Adapted from Graziano, Cooke, et al. (2004).

These findings suggest that either through evolution, experience, or both, the representation of movement in the motor cortex reflects the statistics of the normal movement repertoire. It is not obvious how to optimally arrange something as complex and multidimensional as the movement repertoire onto the two dimensions of the cortical sheet. Solving this problem of dimensionality reduction leads to a theoretical explanation of the complicated, overlapping topography of the cortical motor system. This topic is discussed in the next chapter.

Chapter 10

Dimensionality Reduction as a Theory of Motor Cortex Organization

INTRODUCTION

One way to describe the topography of the cerebral cortex is that "like attracts like." The cortex is organized to maximize nearest neighbor similarity or local continuity (e.g., Durbin and Mitchison, 1990; Kaas and Catania, 2002; Kohonen, 1982; Rosa and Tweedale, 2005; Saarinen and Kohonen, 1985). This principle can explain the large-scale separation of cortex into regions that emphasize different information domains. For example, vision, audition, touch, and movement are roughly separated into large cortical chunks. The same principle can also explain the continuous maps that form within cortical areas.

The reason why the cortex is organized according to proximity is not known, but several plausible explanations can be advanced. One is that it is a side effect of the normal developmental process. During development, axons are guided to their terminations by chemical gradients, and therefore the connectivity from one brain area to another tends to form a topographic continuity (Gierer and Muller, 1995; O'Leary and McLaughlin, 2005). A second possible explanation is that during evolution, information processors that require constant intercommunication tend to be shifted toward each other in cortex to minimize wiring length and thus maximize efficiency. A third possible reason is that neurons that are near each other tend to share more synaptic connections and therefore, during Hebbian learning, become tuned to correlated signals. Probably all of these reasons contribute and interact with each other. For example, it has been suggested that primary cortical maps are hard wired, developing according to genetically programmed chemical gradients, and secondary cortical maps grow in a cascade of Hebbian learning from the primary maps (Rosa and Tweedale, 2005). Whatever the cause for the local smoothness constraint, whether ontogenetic, phylogenetic, or some mixture, the cortex seems to be organized along this principle of like attracts like.

For example, adjacent locations on the retina are mapped to adjacent locations in primary visual cortex in a retinotopic map. Conveniently, both the retina and the cortex are two-dimensional sheets, and therefore the retina can be mapped onto the cortex in a topologically exact fashion. The mapping becomes more complex, however, when a stimulus space that has more than two dimensions is mapped onto the cortical sheet. Optimizing local continuity then becomes a matter of fitting together disparate pieces in the best compromise possible. For example, at the columnar level, the primary visual cortex represents not only the positions of stimuli on the retina but also the orientations of line segments. Durbin and Mitchison (1990) showed that when this three-dimensional stimulus space is reduced onto a two-dimensional sheet, the mathematically optimal configuration in which local continuity is maximized has a pinwheel arrangement that closely resembles the actual arrangement found in the primary visual cortex. It is important to understand that "optimal" in this context is a specific mathematical concept that does not mean "perfect." It refers to the best compromise possible given the task of flattening three stimulus dimensions onto a two-dimensional sheet. By hypothesis, the organization of the cortex is the result of evolution finding the optimal compromise given the resources available.

The finding that the topography of primary visual cortex can be explained by means of a dimensionality reduction greatly supported the case for the principle of optimization of local smoothness. The principle was not merely a verbal tag that summarized the cortical localization of function. It appeared to be able to make mathematically precise predictions about the details of cortical topography. Yet after the use of a dimensionality reduction to model the primary visual cortex, little work was done to determine whether the same principle might explain the topographic details of other cortical areas. The reason why the technique remained limited to the primary visual cortex was probably that the relevant parameter spaces were well known and easily defined. The mathematical problem was circumscribed. In other cortical areas, such as high-order visual areas or motor areas, the parameter spaces were less well known, difficult to define precisely, and much more highly dimensional.

Our stimulation experiments in the monkey seemed to provide some indication of the relevant information dimensions that shape the organization of the motor cortex. These hypothesized dimensions included locations of muscle groups on the body (this aspect of the movement repertoire, if mapped onto the cortex, would tend to produce a somatotopic map of the body), locations in space around the body to which movements are directed (this aspect of the movement repertoire, if mapped onto the cortex, would tend to produce a topographic map of hand space around the body), and the division of the movement repertoire into common, behaviorally useful action types (this aspect of the movement repertoire, if mapped onto the cortex, would tend to produce clusters in cortex that specialize in different common actions). We fed this highly dimensional information domain into a standard dimensionality-reduction engine (Kohonen, 2001) to determine its optimal cortical layout. Although any one of these constraints should have resulted in a simple and orderly map, the simultaneous interaction of the three constraints and the resultant compromise among them produced a complex topography (Aflalo and Graziano, 2006b; Graziano and Aflalo, 2007). The result included blurred maps of the body, gerrymandered borders, gradients, and areas that were separate in some ways and yet fit into a larger map in other ways. The informational space was of such high dimensionality that its reduction onto the cortical sheet did not result in any neatly describable topographic order. Yet this complex topography closely matched the actual pattern observed in the motor cortex of the monkey brain. The model contained subregions that resembled the primary motor cortex, lateral premotor cortex, SMA, FEF, and SEF. The model, therefore, was able to account for the organization of a large sector of cortex comprising about 20% of the cortical mantle. Even an approximate version of the movement repertoire, when reduced onto a two-dimensional surface according to the principle of like attracts like, resulted in a recognizable sketch of the actual cortical topography.

The principle of many movement dimensions competing to form the layout of the motor cortex provides the first plausible theory of motor cortex organization. The classical description of motor cortex as a map of the body is not accurate. The map is blurred, overlapping, and partially repeated in ways that are difficult to pin down (see Chapter 3). The common modern description of motor cortex as a mosaic of separate areas with different functions is, first, an oversimplification of a blurred and subtle pattern, and second, merely a description without an underlying explanation. The present theory that the movement repertoire is reduced onto the cortical sheet seems finally to provide a working explanation of the layout of motor cortex. The model described below is approximate, relying on rough descriptions of the movement repertoire and of the shape of the cortical sheet. Therefore the model itself is unlikely to be correct in all its details. However, the model comes close enough to reality to suggest that the underlying principle is probably the correct one.

METHOD: QUALITATIVE DESCRIPTION

Three types of movement dimension were used to inform the model: somatotopic, ethological action category, and spatial. Each type by itself was of low enough dimensionality that it could have been mapped onto the cortical sheet in a simple and orderly map. The three together, however, presented a more complex optimality problem. To optimize one type of map would be to scramble the other two types of maps. The global optimum therefore required a compromise among the three potential maps. In this sense, the three potential maps competed with each other for the organization of the cortical sheet.

Somatotopy

In the model we defined a set of twelve body parts that could be mapped across the cortical surface. We assigned the model an initial somatotopic organization based on the map of the lateral motor cortex published by Woolsey et al. (1952). The initial state of the model is shown in Figure 10-1. It is important to note that this use of the Woolsey map to initialize the model was not strictly necessary. If initialized totally randomly, during the optimization process the map could nonetheless form a somatotopic organization. The usefulness of initializing the model with the Woolsey map was that it forced the

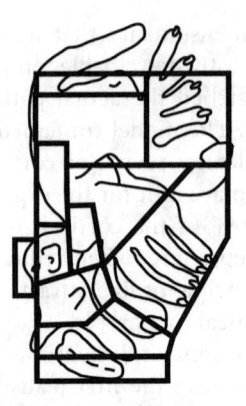

Figure 10-1 The initial state of the map model. The map of the monkey body in the lateral motor cortex according to Woolsey et al. (1952) is shown, with an overlay showing the simplified, blocked arrangement of 12 body parts defined as the initial state of the motor cortex model. This map of Woolsey et al. served as a "trellis" for the model, ensuring that during the optimization process the model formed a somatotopic map oriented in the correct fashion.

somatotopic organization to have a particular orientation, with the head down, the feet up, the distal muscles emphasized to the right, and the proximal muscles emphasized to the left. Thus the importance of the Woolsey map was something like a trellis to orient the model correctly. As described below, the details of the Woolsey map disappeared as the model gradually shaped itself to optimize the simultaneous constraints. Only the general orientation of the somatotopic map remained.

Ethologically Relevant Action Category

In the model, in addition to defining a set of body parts that could be mapped across the cortical surface, we also defined a set of eight action categories. In seeking local continuity, the model tended to form clusters for each action category. These action categories included hand-to-mouth movements, manipulation of objects in central space, reaching to grasp, defensive movements (including arm withdrawal and facial defensive movements), chewing, bracing the hand in lower space, complex locomotion such as climbing, and gaze shifts (head and eye movements). These action categories constitute the main part of a monkey's normal behavioral repertoire. They are also all readily evoked by stimulation in motor and oculomotor cortical areas. Each action category combined more than one body part. A hand-to-mouth action, for example, combined the hand, arm, neck, jaw, and lips.

Hand Location

In the model, those movements that involved the arm were also assigned a hand position in space. During optimization, the map sought continuity in

10. Dimensionality Reduction as a Theory of Motor Cortex Organization 155

this representation of hand location. Any possible hand location map, however, was necessarily constrained by the simultaneous mapping of action categories. The reason is that each action category was associated with a characteristic set of hand locations (Figure 10-2). For example, hand-to-mouth movements were associated with hand locations in a small region of space around the mouth, climbing-like movements were associated with hand locations generally distant from the body and distributed in the frontal and lateral space, and so on.

Optimization

The cortical map was optimized according to the method of Kohonen (2001). The Kohenen method is a standard tool for solving the problem of dimensionality

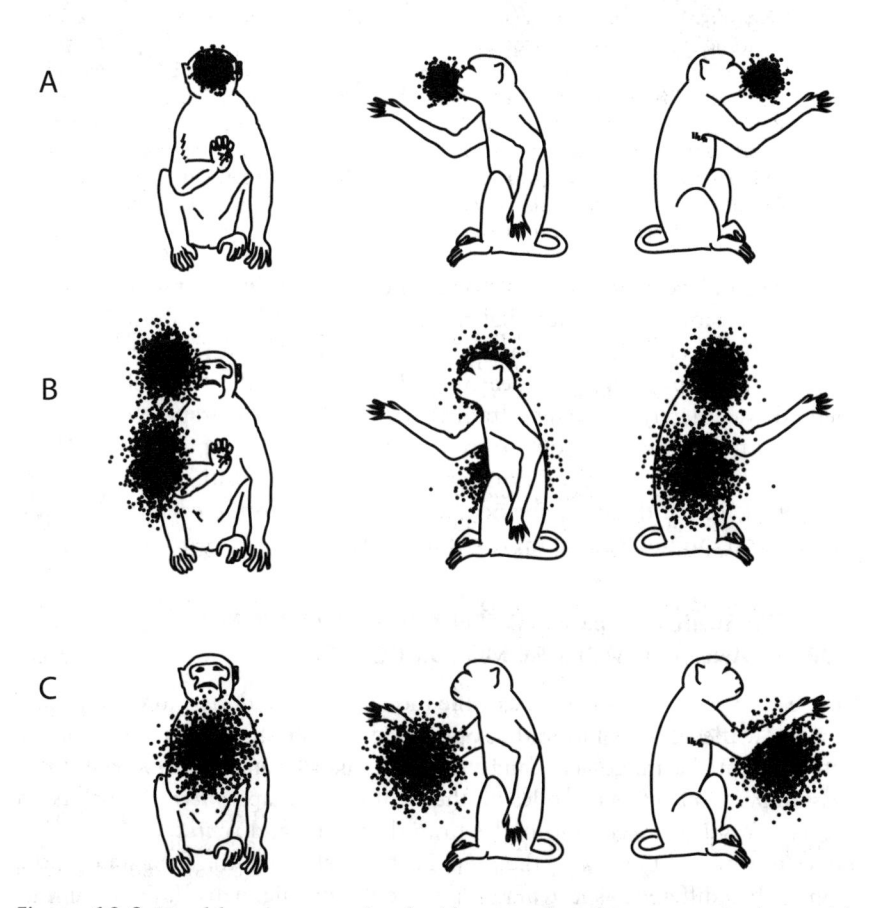

Figure 10-2 Hand locations associated with categories of movement in the model. Three views of a schematized monkey showing the distribution of hand locations assigned to hand-to-mouth movements (A), defense (B), manipulation of objects (C). reaching (D), and climbing (E).

(Continued)

D

E

Figure 10-2 cont'd

reduction, or the problem of representing a multidimensional space on a lower dimensional space such that neighbor relationships are optimized. For the present purpose, the Kohonen method was not meant to model the specific neuronal interactions or learning algorithms of the brain. Rather the method was merely an analytic tool that optimized topographic continuity. Details of the implementation are given in Aflalo and Graziano (2006b). These details of how somatotopy, action category, and hand position were encoded in a set of formal dimensions and then flattened onto the cortical sheet are not necessary to understand the general points discussed below.

RESULTS: SIMILARITIES BETWEEN THE MODEL MOTOR CORTEX AND THE MONKEY MOTOR CORTEX

Figure 10-3 shows the final state of the model, after the dimensionality reduction had settled on a solution that optimized local continuity. For consistency with the spatial arrangement found in the monkey brain, in the following discussion *posterior* refers to nodes on the right of the map, *anterior* to nodes on the left, *dorsal* and *medial* to nodes toward the top, and *ventral* and *lateral* to nodes toward the bottom of the map. Each panel shows the final state of the map with a different aspect of the representation highlighted. For example, panel A shows the representation of the tongue, mainly in the ventral part of the map. Panels B–L show the representations of the other body parts. Panel M shows the representations of the eight explicitly defined ethological action

10. Dimensionality Reduction as a Theory of Motor Cortex Organization 157

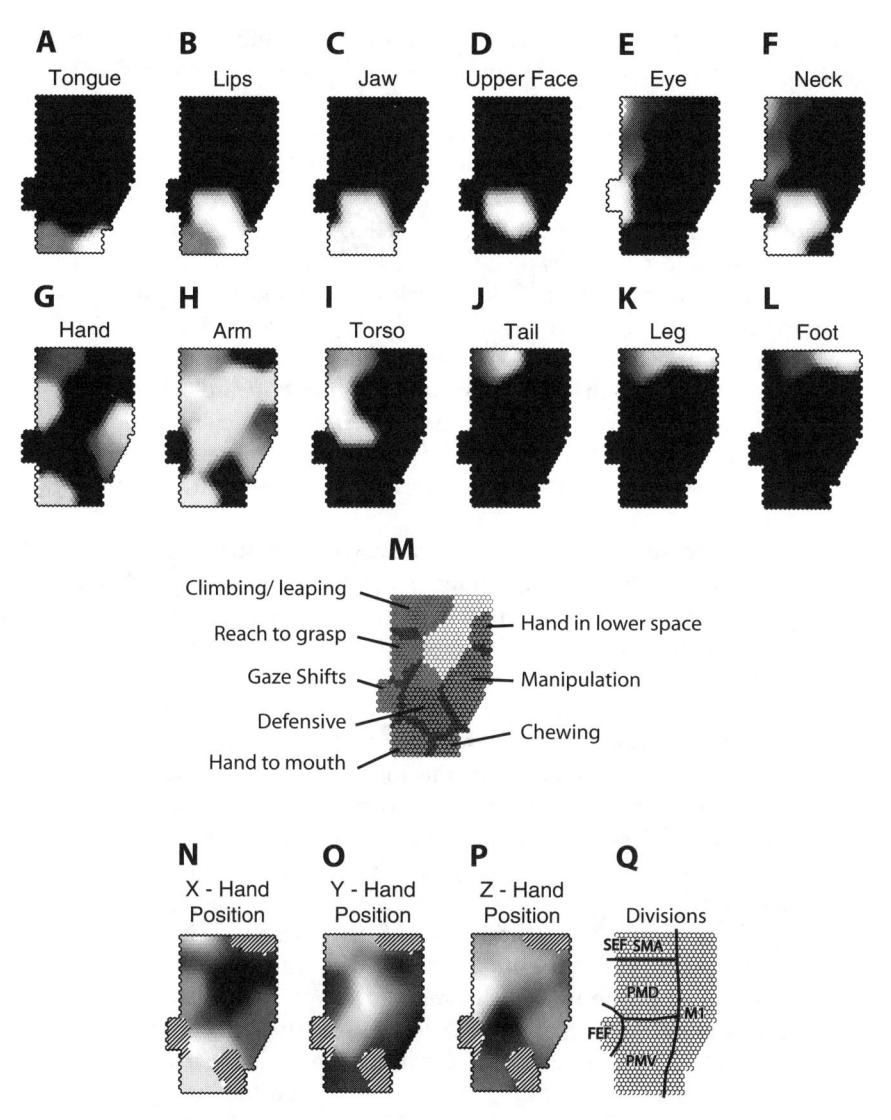

Figure 10-3 Final state of the self-organizing map model. **A–L.** Representations of the 12 body parts after map reorganization. White=map locations in which the body part is more strongly represented. **M.** Arrangement of the eight ethological categories of movement after reorganization. **N–P.** Maps of hand location after reorganization. Only those nodes that had a nonzero magnitude of arm representation are shaded because only these nodes had a defined hand position. Nodes with no arm representation are cross hatched. X=hand height, white=greater height; Y=lateral location of hand, white=more lateral locations; Z=distance of hand from body along line of sight, white=more distant locations. **Q.** Some common divisions of the monkey motor cortex drawn onto the map model.

categories. Panels N–P show the representations of hand position across the map model. Panel Q shows a hypothetical demarcation on the model of some commonly accepted divisions in the monkey motor cortex.

The topography generated by the artificial model of motor cortex is similar to the actual motor cortex of the monkey in the following ways.

1. As a result of the dimensionality reduction, the somatotopy is a blurred one with considerable overlap among adjacent body-part representations, similar to the actual maps obtained in physiological experiments (e.g., Donoghue et al., 1992; Gould et al., 1986; Park et al., 2004; Park et al., 2001; Sessle and Wiesendanger, 1982). The reason for the somatotopic overlap is straightforward. Most of the movements incorporated into the model involved combinations of body parts. Therefore, in developing representations of those actions, the map was forced to develop overlapping representations of body parts.
2. The model developed a distinction between a posterior strip of the map and an anterior strip. Along the posterior strip (the right edge of the array), a relatively discrete progression can be seen. This progression includes a mouth representation at the bottom, then a region that emphasizes the hand but also weakly represents the arm, then a region that emphasizes the arm but also weakly represents the hand, then a region that represents the foot and leg. A classical motor somatotopy is displayed. Along the anterior strip of the map (the left edge of the array), the somatotopy is much more overlapping and fractured, and a classical motor somatotopy is not as evident, consistent with the overlapping topography typical of the monkey premotor cortex.

    The reason for this trend in the self-organizing map is clear. Some of the movements in our model required coordination among major segments of the body. These movements involved the axial musculature because the trunk and neck form the connecting links between different body segments. The initial somatotopy was arranged with the axial musculature in an anterior region and the distal musculature in a posterior region. As a result, during map optimization, the complex movements that link more than one body segment gravitated to the anterior regions of the map. For example, reaching involved not only the arm and hand but also the torso and thus emerged in an anterior location; hand-to-mouth movements involved the neck to coordinate between the arm and the mouth, and thus emerged in an anterior location; climbing-like movements involved the neck and torso as the connecting links between head, arms, and legs, and therefore emerged in an anterior location. Thus in our model, in its final state, one can distinguish a posterior strip that is "primary-like" in that it contains a relatively discrete somatotopy, representing body segments in a partially separate manner, and an anterior strip that is "premotor-like" in that it contains a more integrated, blurred somatotopy and represents movements of greater inter-segment complexity.

3. The model developed a blurred, secondary map of the body that resembled the SMA body map found in the monkey cortex (e.g., Macpherson, Marangoz, et al., 1982; Mitz and Wise, 1987; Woolsey et al., 1952). This secondary map in the model was located along the medial edge, progressing from a representation of the foot in a posterior location, through a representation of the trunk and arm, to a representation of the head and eye in an anterior location. The reason for the emergence of this secondary somatotopy in the model is clear. It is a mapping of the action category related to complex locomotion. Locomotion in a complex environment strewn with obstacles, in which the hands and feet need to be placed on disparate opportune surfaces, includes all limbs, the head, the eye, the torso, and the tail as a balancing device. Not all body parts are moved simultaneously; instead the actions form an overlapping distribution, some movements weighted more toward the upper body and some weighted more toward the lower body. This highly overlapping distribution of movements, incorporated into the model, resulted in an overlapping map of the body that emerged adjacent to the original leg and foot representation.

4. The hand representation became divided into three main regions (Figure 10-3G). One hand representation was located in the posterior part of the array, as if corresponding to the primary motor hand area; the second hand representation was located in an anterior region within the dorsal half of the array, as if corresponding to the dorsal premotor hand area; and the third hand representation was located in an anterior region at the ventral edge of the array, as if corresponding to the ventral premotor hand area. These three hand areas also resemble the three lateral hand areas described by Dum and Strick (2005) on the basis of projections from cortex to the spinal cord. The reason why the model developed three distinct hand areas is that it was trained on three distinct categories of action that emphasized the hand: manipulation in central space (represented in the posterior region), reaching to grasp (in the dorsal anterior region), and hand-to-mouth movement (represented in the ventral anterior region).

5. The posterior hand representation in the model was partially surrounded by a region of greater arm representation. This can be seen by comparing the posterior hand representation shown as a bright region on the posterior edge of the map in Figure 10-3G to the dark spot at the same location, with relatively little arm representation, in Figure 10-3H. The pattern is therefore of a central region on the pos-

terior edge of the map that emphasizes the hand with relatively less involvement of the arm, and a surrounding half ring of cortex that emphasizes the arm with relatively less involvement of the hand. This core and surround organization resembles the organization found in the monkey primary motor cortex (Kwan et al., 1978; Park et al., 2001). The reason for this organization in the model is that there is a range of actions involving different relative contributions of the arm and hand. The actions that emphasize the hand tend to cluster, as the map seeks to optimize nearest neighbor relationships. The actions that emphasize the arm, however, have a greater diversity, including a range of arm positions in space around the animal, and therefore do not cluster to the same extent. The cortex just dorsal to the core hand area emphasizes arm locations in lower space. The cortex just ventral to the core hand area emphasizes arm locations in more elevated space. The core-surround organization in the model is therefore a result of a complex interaction among several mapping requisites.

6. The eight ethological categories of movement became focused into eight cortical zones that were relatively discrete, with minimal overlap (Figure 10-3M). The topographic arrangement of the zones in the self-organizing map resembled the arrangement observed in the actual monkey brain. This arrangement of ethological zones resulted from the initial somatotopy and the subsequent attempt of the model to optimize nearest neighbor relationships. For example, the hand-to-mouth movements converged on a ventral location where the mouth, hand, and arm representations could most easily develop a region of overlap. The climbing movements converged on a dorsal location where the arm, leg, and torso representations could develop a region of overlap. The reaching movements converged on a region where the arm, hand, and torso could most easily develop a region of overlap. In this manner, the relative position of these action zones on the cortex was highly constrained. The apparent "hole" in the map in Figure 10-3M was filled with general hand and arm movements that were not included in the eight labeled ethological categories.

7. The defensive zone developed an internal topography in which arm-related defensive movements were represented in the dorsal part of the defensive zone and purely face-related defensive movements were represented in the ventral part of the defensive zone (compare Figure 10-3D, H, and M). This arrangement emerged because, in the initial somatotopy, the face was represented in a more ventral location than the arm, biasing the final configuration. The arrangement matched the results from the monkey cortex (e.g., Graziano et al., 1997a; Graziano, Taylor, et al., 2002). In the corresponding zone in the monkey cortex, some neurons have tactile responses on the arm and visual responses near the arm, and stimulation of these neurons evokes arm retrac-

10. Dimensionality Reduction as a Theory of Motor Cortex Organization   161

tion. These neurons tend to be located in the dorsal part of the defensive zone. Other neurons have tactile responses on the face and visual responses near the face, and stimulation of these neurons evokes face-related defensive movements. These neurons tend to be located in the ventral part of the defensive zone. The model therefore correctly captured this detail of the monkey motor cortex.

8. The model developed noisy maps of hand location that approximated the findings in the monkey motor cortex. The height of the hand (Figure 10-3N) was most consistently mapped, with upper hand positions in a ventral location in the map and lower hand positions in a dorsal location. A dorsal, anterior region of the map, overlapping the representation of climbing-like movements, represented a range of hand locations again roughly matching our findings in the monkey brain. The lateral position of the hand (Figure 10-3O) was less clearly ordered, and the forward distance of the hand along the line of sight (Figure 10-3P) showed little consistent topography, again consistent with the findings from monkey cortex. The crosshatched regions of these maps indicate regions where no arm movement was represented and therefore hand position was undefined.

9. The model developed two hot spots for combined eye and head movement, resembling the locations of FEF and SEF in the monkey brain (Figure 10-3E). The FEF-like area was in an anterior, lateral location. This area resulted from the initial somatotopic arrangement in which the eye was represented in that location. The SEF-like area was in an anterior, medial location, in the most anterior part of the SMA-like region of the map. This SEF-like area developed because of the inclusion of gaze shifts in the complex locomotor action category. In these respects the model converged on an arrangement essentially identical to that in the monkey brain.

The model did not incorporate any dimensions related to the specific vectors of eye movements. Nonetheless, one feature of topography can be discerned in the FEF-like area in the model. The model placed the pure eye-movement representation in the ventral part of the FEF, and the combination of eye movement and neck movement in the dorsal part of the FEF. This can be seen by comparing the distribution of the eye and the neck representations in Figure 10-3E and F. A topography of this type is also present in the actual monkey FEF. Indeed the main organizational feature of the FEF in the monkey, and the only topographic feature that has been consistently found, is a tendency for long amplitude gaze shifts that require eye and head movements to be represented in dorsal FEF, and small amplitude saccades that do not require head movements to be represented in ventral FEF (Bruce et al., 1985; Knight and Fuchs, 2007). The model therefore correctly reconstructed this detail of topography within the FEF.

In a similar manner, within the SEF-like area in the model, combined eye and neck movements were represented preferentially in the anterior part. In the actual monkey brain, long saccades that recruit the eye and the head are preferentially represented in the anterior part of SEF (Chen and Walton, 2005; Tehovnik and Lee, 1993).

Thus the model correctly reconstructed the essential features of the topography in both eye movement areas.

10. Although the representation of eye movement in the model became focused on an FEF-like and SEF-like area, some representation of eye movement also developed in the cortex between these two areas, in the PMD-like region of the map. In the actual monkey brain, eye movement is indeed represented to some degree in PMD, and this eye movement representation is stronger in the more anterior part of PMD (Boussaoud, 1995; Bruce et al., 1985; Fujii et al., 2000). The reason why the model developed an eye representation in this region is because of the representation of reaching in the same region of the map. It has been reported that reaching to grasp an object and gaze movements are often integrated (Mennie et al., 2007). Therefore in the definition of movements supplied to the model, the reaching-to-grasp category was composed of a range of arm and hand movements some of which were associated with eye movement.

11. The organization of the cortical motor areas is essentially consistent among monkeys. An important question is whether the topography produced by the present model, that closely matches many of the features of the real topography, is robust or whether it changes radically with a small change in the input parameters. We therefore tested variants of the model including alterations in the size and shape of the cortical sheet used as a basis for the model, variations in the proportions of different movements supplied to the model, modifications to the hand locations assigned to different movements, and variations in the proportions of different body parts combined within each movement. Every parameter that was used as input to the model was varied, while preserving the same general description of the movement space. These changes in the information used to seed the model resulted in small changes in the final result in the exact size and exact locations of functional regions. The overall pattern, however, remained robust. The arrangement of functional zones in the model converged on more or less the same optimal configuration. This testing of the model by changing the parameters is described in greater detail in Aflalo and Graziano (2006b).

12. Previous studies have shown that the topography in motor cortex can change with experience or with lesions to parts of motor cortex (e.g., Classen et al., 1998; Karni et al., 1995; Kleim et al., 1998; Nudo and Milliken, 1996; Nudo et al., 1996; Pascual-Leone et al., 1995).

10. Dimensionality Reduction as a Theory of Motor Cortex Organization

We tested the model to determine if it could explain these types of plasticity.

In one test we allowed the model to optimize as above, producing a cortical organization as in Figure 10-3. We then altered the statistics of the movement space, doubling the proportion of manipulation actions, and trained the model further. As a result, the manipulation zone in the map model expanded. In this simulation, therefore, greater "practice" on one type of behavior resulted in an expansion of the corresponding representation.

In a second test, we again trained the model to produce a cortical organization as in Figure 10-3. We then "lesioned" the manipulation zone by removing most of the nodes in that region of cortex, leaving a hole in the map model. On further training, the model reorganized such that the small remnant of the manipulation zone expanded into neighboring cortex. In this simulation, a lesion to one part of the map with subsequent practice on a normal movement repertoire caused neighboring areas of cortex to take over the lost representation.

These tests suggest that some of the changes in the motor map that are known to occur with practice or with lesions can be understood in the context of the present model. If it is true that the movement repertoire is rendered onto the two-dimensional sheet of the motor cortex, then changing the statistics of the movement repertoire, or making a physical hole in the motor cortex, will change the rendering in predictable ways.

LIMITATIONS OF THE MODEL

1. A set of little understood motor areas in the monkey cortex lies in the cingulate sulcus on the medial wall of the hemisphere (Dum and Strick, 1991). These areas are absent from the model. The reason is that there is not yet any known functionality for those areas to supply to the model. If the cingulate motor areas were electrically stimulated, what movements would be produced, and would those movements be recognizably part of the animal's normal repertoire? This experiment has not yet been done, nor have single neuron experiments explored the specific functions of those areas. The model does not, of course, create or discover functionality. Its intended goal is to explain why known functions are arranged as they are on the cortex.
2. In the actual monkey motor cortex, the oral representation is larger and extends more ventrally than the corresponding representation in the model. The probable reason for the model's inaccuracy is its

impoverished description of oral movements. In reality, in addition to chewing, oral behaviors include spatially precise and complex movements of the tongue within the mouth, swallowing, vocalization, food pouch storage, and probably other behaviors. A more inclusive description of the motor repertoire in this case would be expected to lead to a larger and more diverse mouth and tongue representation in the model. Comparing the model to the reality, however, would be difficult because little is known about the organization of the mouth and tongue representation in motor cortex (but see Huang et al., 1989).

3. The model contains a coarse, first-order description of the movement repertoire. For example, in the model, the forelimb is divided into a hand and an arm, ignoring the details of fingers, wrist, forearm, elbow, and shoulder. Hand position is incorporated into the model, with different types of actions having different distributions of hand positions, but the model does not include the complex, multijoint arm postures normally associated with those actions. Each action type in the model is uniform, whereas in reality there are many variants of each action type. This rough description of large segments of the movement repertoire appears to be sufficient to capture the large-scale organization of the cortical motor areas, but it ignores the possibility of a more fine-grained structure such as at the columnar level for the representation of movement details. With more specific information about the statistics of a monkey's movement repertoire, it may be possible to extend the model in this direction to determine if it can correctly predict columnar organization.

4. The hypothesis tested here is that information is arranged across the actual monkey cortex according to the same like-attracts-like optimality principle used by the model. The model, however, does not address how that optimization occurs in the brain. The cortex presumably finds this optimal organization through a combination of evolution and experience-dependant fine-tuning. Through this process, a region of the cortex comes to emphasize eye movements, another region emphasizes reaching, and so on. These cortical zones take on the cytoarchitecture and connections useful for their specific information domains. The major cytoarchitectural and connectional properties that define cortical zones are presumably the result of optimization through evolution. Smaller fine-tuning of cytoarchitecture and connections is presumably possible through experience. These issues of evolution versus learning, and of cytoarchitecture and connections, however, are not directly addressed by the model. Instead the model directly addresses only the mapping of information across the cortex and assumes that the physical properties of cortex, such as cytoarchitecture and connections, follow the functionality.

5. The model as it stands does not encompass all cortical motor areas. Rather it encompasses a set of areas that directly control the movement repertoire through output to subcortical motor nuclei and the spinal cord. Other cortical regions such as the parietal motor areas and rostral premotor areas, outside the perimeter of the present model, may play other roles in the control of movement. Some of these possible cortical interrelationships are discussed in Chapter 4.

Chapter 11

Feedback Remapping and the Cortical-Spinal-Muscular System

INTRODUCTION

One of our current goals is to construct a model of the cortical-spinal-muscle system (the CSM model) that is as true as we can get to the actual primate neuronal and muscular architecture, and to study the manner in which this model system can or cannot learn complex movements typical of a monkey's normal repertoire. Other neural network models of the control of the arm have been published (e.g., Maier et al., 2005; Todorov, 2000). Our purpose is to build on this previous work to develop a model that is a closer approximation to the actual system. The model arm (a simulated rather than a physical arm) includes a realistic skeletal structure of the arm and shoulder with 13 degrees of joint rotation and a set of muscles with realistic force-generating properties and insertion points. The model control system includes as much of the cortical-spinal architecture as we can accommodate. The movements under study include reaching, defensive movements, hand-to-mouth movements, and other behaviors that are typical of the natural repertoire and that can be evoked by stimulation of the motor cortex.

The main goal of building the model is to understand how the cortex, spinal cord, and muscles operating together in an integrated system can solve problems in movement control. The project is ongoing. As the CSM model progresses we find that some of the most complex and difficult conundrums of motor control give way when confronted with two basic features of the neuronal architecture.

One fundamental feature of the neuronal system is the divergence of connectivity through the network. One cortical neuron, when active, can alter the activity of a large set of spinal interneurons. One spinal interneuron can influence a large set of muscles. This connectional divergence leads directly to the formation of what has been termed "muscle synergies" (Giszter et al., 1993). A muscle synergy is a stored pattern of relative activation levels across a group of muscles. For example, a single interneuron in the spinal cord can influence a set of muscles, supplying activity to the muscles in a fixed ratio. Useful muscle synergies can be learned by the divergent connectivity in the network and called upon when needed. In frogs, cats, and humans, complex movement can be decomposed into a relatively small number of muscle synergies that appear to act as building blocks and that can be combined linearly in a time-varying fashion (D'Avella et al., 2003; Ting and Macpherson, 2005; Torres-Oviedo and Ting, 2007; Tresch et al., 1999).

A second fundamental feature of the neuronal system is that feedback about the state of the periphery modulates the entire network at every level and therefore alters the relative strength of pathways through that network. If the network receives reliable feedback about a specific movement variable, then it can learn to control that variable. Thus it is not necessary to hypothesize that the controller is fundamentally a postural one, or a directional one, or a controller of muscle force or of muscle activity. Instead, a range of behaviorally useful variables can be controlled by this network in any combination.

These two basic properties of the neuronal architecture, the linking of muscles into useful sets by means of divergent connectivity, and the use of feedback to regulate task-relevant variables, form a powerful combination that allows the network to learn the control of complex movement. We termed the combination of these two properties "feedback remapping." The first property corresponds to the mapping from each neuron in cortex to a spatial pattern of activation and inhibition across a set of muscles. The second property refers to the manner in which feedback about the periphery continuously modulates the cortical and spinal circuitry, thereby altering that mapping through time.

The details of the CSM model are outside the scope of this chapter. We are still studying the model, and the results will be published elsewhere. The purpose of this chapter is to describe in a step-by-step fashion some of the basic building blocks of feedback loops and muscle synergies, to reach, by the end of the chapter, a qualitative explanation of the movements evoked by electrical stimulation of motor cortex. As described in the previous chapters, electrical stimulation of sites in the motor cortex can evoke movements of great complexity, recruiting many muscles of many body parts. The movements resemble real actions in the behavioral repertoire of the animal. Yet the movements evoked by stimulation are not identical to natural movements. They are lacking in some aspects of control, in nuance, and especially in the temporal dynamics of muscle activity. How can stimulation in the primary motor cortex, only a few synapses from the muscles, result in such complex, behaviorally meaningful actions, and given that it does so, why do the movements deviate from natural movements in the ways that they do?

In a traditional view, the motor cortex operates by a set of cables that project downward, with a relay in the spinal cord, to the muscles. Activity at a site in motor cortex results in activity in the corresponding muscle or group of related muscles. A map of this simplistic type is unable to account for the complex movements evoked by electrical stimulation in our experiments. In fact, we encountered considerable skepticism of our work from scientists who insisted that, the motor cortex being a simple map of muscles, our results were impossible. The truth is, however, that the motor cortex is not a simple map of muscles. The circuitry is far more complex than a set of cables from the cortex to the muscles. Once the complexities of this circuitry are clearly spelled out, it is quite straightforward to explain the general properties of the stimulation-evoked movements.

The following sections first describe some previously proposed models of the neuronal control of movement. These models include the λ model and the hypothesis of muscle synergies. The chapter then discusses the usefulness of higher-order sources of information about the periphery such as visual feedback and internally generated models of the body. The chapter ends with a schematic version of the cortical-spinal-muscle system and a discussion of how electrical stimulation in cortex is likely to affect this circuitry and result in complex movement.

FEEDBACK CONTROL AND THE λ MODEL

One of the earliest neurally inspired models of feedback control is the λ model of Feldman and colleagues (Asatryan and Feldman, 1965; Feldman, 1966; for a recent review, see Feldman and Latash, 2005). In the λ model, stretch receptors in muscles provide feedback to α motor neurons. A simple, schematic version of the λ model is shown in Figure 11-1. In this schematic, for simplicity, a muscle is controlled by a single α motor neuron, and the length of the muscle is detected by a single stretch receptor. In reality, of course, a pool of α motor neurons would control the muscle, and the muscle would contain many stretch receptors. However, this simplification does not change the underlying concepts of the model.

The activity in the α motor neuron is determined by at least two signal sources. First, the α motor neuron receives a descending signal from a higher-order source such as a spinal interneuron. Second, feedback from the stretch receptor also provides input to the α motor neuron. These two signals combine. Whether or not the α motor neuron passes the threshold for generating action potentials, and how high above threshold it is driven, depends on that combination of signals. For this reason, a specific, steady-state signal from above does not translate into a specific level of muscle activity. Instead it effectively sets a threshold length for the muscle, termed "λ." If the muscle is stretched beyond λ, then the stretch receptor provides enough excitatory feedback to drive the α motor neuron above its firing threshold, thereby generating a restoring force in the muscle. The farther beyond λ the muscle is stretched, the farther above threshold the α motor neuron is driven, and therefore the greater the restoring activity generated in the muscle.

In principle, the threshold length for the muscle can be adjusted in two ways, both implicit in Figure 11-1. First, the direct descending signal to the α motor neuron helps bring it closer to its firing threshold. Second, the descending signal to the γ motor neuron adjusts the stretch receptor itself, changing its sensitivity, thereby changing the gain on the feedback signal. Through both routes, the descending signal sets the threshold length beyond which the muscle becomes active and generates a restoring force.

By controlling muscle length, the λ model can control joint angle. Consider a joint that is crossed by two muscles, a flexor and an extensor. By setting the threshold lengths of these two muscles, the system sets the desired joint angle. If the joint is flexed beyond the desired angle, then the extensor muscle is stretched beyond its threshold length and generates a restoring force. If the joint is extended beyond the desired angle, then the flexor muscle is stretched

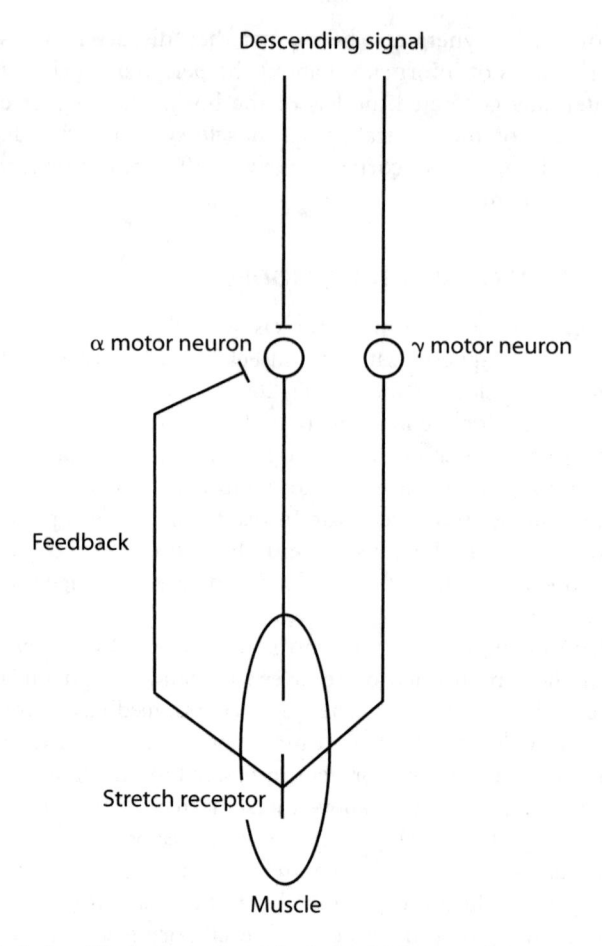

Figure 11-1 A simplified schematic of the λ model for muscle control. The descending signal effectively sets a threshold length, λ. If the muscle is stretched beyond λ, the feedback from stretch receptors excites the α motor neurons above their firing threshold, generating a restoring force in the muscle. In this schematic, only one α motor neuron, one stretch receptor, and one γ motor neuron is shown. In reality a muscle would contain many of each.

beyond its threshold length and generates a restoring force. In this manner the descending signal can specify a desired joint angle that is defended by the feedback circuitry. The descending signal can also specify a series of desired joint angles through time.

One limitation of the λ model is its sole reliance on muscle stretch as the only explicit control variable. The spinal cord and motor cortex receive proprioceptive and visual feedback about a range of variables including joint speed, muscle tension, pressure on the skin, hand trajectory, limb posture, and so on. The system can potentially control these many variables through feedback

11. Feedback Remapping and the Cortical-Spinal-Muscular System 171

loops. In this sense the λ model may be a correct but limited description of the control of one kind of variable by means of one kind of feedback.

A second limitation of the λ model is that the stretch reflex may be too weak by itself to allow the joint to closely track the instructed angle (e.g., Lan and Crago, 1994). Again, the λ model may be a valid description of one part of a larger control system. The feedback loop that is at the heart of the λ model, the monosynaptic stretch reflex, is only one example in an extensive system of feedback loops. The weak gain on the stretch reflex may reflect the fact that it is working in combination with other feedback loops.

FEED-FORWARD CONTROL AND MUSCLE SYNERGIES

The motor system is organized partly as a diverging, branching tree of connectivity from each higher-order neuron to many muscles. This divergence in the connectivity allows a certain kind of feed-forward control. Each higher-order neuron instructs a set of muscles to become active in a specific ratio or "muscle synergy." The role of this type of control signal was studied in detail by Bizzi and colleagues (e.g., D'Avella et al., 2003; Giszter et al., 1993; Tresch et al., 1999). Their hypothesis is diagrammed schematically in Figure 11-2. This figure shows two example interneurons in the spinal cord. Each interneuron influences many muscles. If interneuron 1 produces muscle synergy 1, and interneuron 2 produces muscle synergy 2, then a higher-order signal activating those two interneurons can in principle produce any linear combination of the two synergies. The hypothesis of muscle synergies is therefore a hypothesis

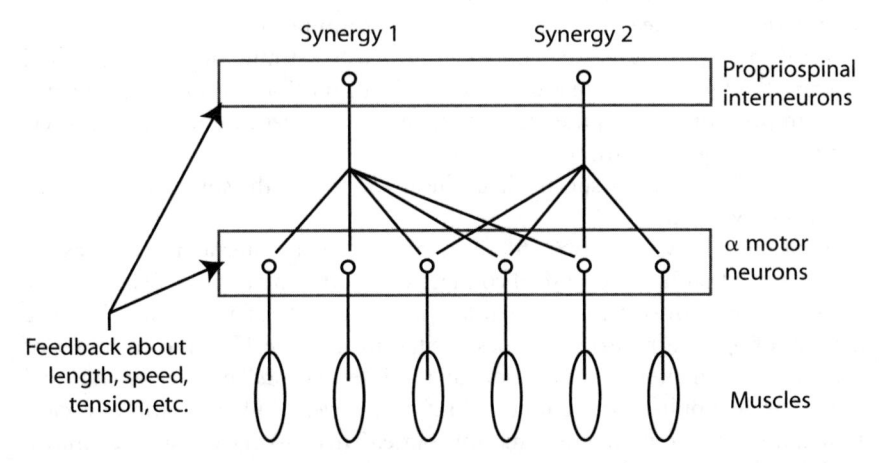

Figure 11-2 One possible way that muscle synergies might be built into the spinal circuitry. A propriospinal interneuron in the spinal cord connects to the α motor neurons of many muscles, with a fixed set of connection weights. Activity of that interneuron therefore supplies excitation in a fixed ratio to the set of muscles. Sensory feedback can modulate the activity of the α motor neurons and the propriospinal interneurons.

about the basis set of combinable elements. The elements, in this hypothesis, are not individual muscles but useful patterns of activity across muscles.

The hypothesis of muscle synergies was developed from a set of experiments on the frog spinal cord. Giszter et al. (1993) electrically stimulated the interneuron layers in the spinal cord of frogs and measured the effect on the hind leg. The stimulation did not directly activate the α motor neurons of the leg muscles. The latency from stimulation onset to muscle activity onset confirmed that the effect was a secondary one in which the electrode activated interneurons that in turn activated α motor neurons.

To assess the effect of stimulation, the experimenters measured the activity of muscles in the frog's leg. When stimulation was applied to a site in the spinal cord, the leg muscles became active. Each stimulation site evoked a specific ratio of activity among the leg muscles. Greater stimulating current produced greater overall activity but still produced a similar ratio of activity among the muscles.

To study the effect of this pattern of muscle activity on leg movement, the experimenters held the frog's leg in different positions with a strain gauge attached to the ankle. For each ankle position, spinal stimulation was applied and the magnitude and direction of the evoked force on the ankle was measured. In this manner a "force field" was plotted, consisting of a force vector measured at each location in space.

One type of force field, obtained from some stimulation sites, was a convergent one. In these cases, the stimulation appeared to drive the foot toward a specific location in space. This point of convergence could be described as the equilibrium position of the limb, or the point at which the muscle forces were in balance. When the foot was released from its holder and free to move, stimulation did indeed drive it to that equilibrium position.

A second type of force field, obtained from other stimulation sites, appeared to drive the foot toward an edge of the range of motion. In this case, an equilibrium position is not quite the correct description because the active muscle forces drove the limb toward an extremum.

These results led to a specific hypothesis about how the spinal interneurons control movement.

First, muscle synergies are stored by propriospinal interneurons, a class of interneuron that has widespread connectivity in the spinal cord. Each of these interneurons connects to the α motor neuron pools of many muscles. The activity of an interneuron provides excitation in a specific ratio to its set of muscles. The greater the excitation of the interneuron, the greater the total excitation it supplies to its connected set of muscles, while the ratio of excitation among the muscles remains the same. This linking of muscles into a functional set is a muscle synergy.

Second, only a small number of muscle synergies are stored in the spinal cord. In the case of the frog, about half a dozen were discovered through electrical stimulation. Therefore there is some redundancy in which many spinal interneurons that are spatially near each other have similar connectivity onto the α motor neurons. The result of this similarity among adjacent interneurons is that electrical stimulation of sites anywhere within a zone in the spinal gray matter will evoke essentially the same muscle synergy. Only when the stimulating electrode is moved to a functionally different zone in the spinal gray will the muscle synergy change. In Figure 11-2, each synergy is represented schematically by a single interneuron, but in reality each synergy is probably represented by a set of interneurons with similar properties.

Third, these muscle synergies are not limited to the control of one joint. The spinal interneurons integrate among joints. A single muscle synergy can affect the entire frog leg and perhaps muscles in the torso as well.

Fourth, each muscle synergy tends to produce a goal configuration of the limb. Some of these configurations correspond to an equilibrium position of the limb, and others correspond to an extreme position of the limb.

Fifth, it is not correct that every possible desired position or configuration of the limb is separately coded in the spinal circuitry as a distinct, stored muscle synergy. It is evidently not the case that the limb is moved to a specific position by dialing in a new equilibrium position. Instead, because only a small number of muscle synergies were found, a small set of underlying synergies must act as a basis set, called up in different time-varying combinations to control more complex trajectories. Exactly how these synergies are recruited and combined by higher-order signals is not addressed by the frog experiment. One hypothesis is that, in primates, the motor cortex represents a higher-level mechanism that recruits the spinally organized muscle synergies. In this view, in effect each cortical neuron recruits a large and complex muscle synergy that, itself, is composed of smaller spinally organized synergies.

This view of muscle synergies stored within the system by means of divergent connectivity, and combined linearly as a basis set for more complex movement, has been supported by a line of subsequent experiments. Similar muscle synergies were evoked by stimulating the spinal cord of the rat, demonstrating that this aspect of motor control is not limited to amphibians but is present in the mammalian nervous system as well (Tresch and Bizzi, 1999). In addition, in frogs, cats, and humans, the varied patterns of muscle activity during the normal movement of the leg can be explained as time-varying linear sums of a small number of underlying muscle synergies (D'Avella et al., 2003; Ting and Macpherson, 2005; Torres-Oviedo and Ting, 2007; Tresch et al., 1999).

INTEGRATING THE λ MODEL WITH MUSCLE SYNERGIES

The hypothesis of muscle synergies described above involves feed-forward control. Each spinal interneuron sends a feed-forward signal to a set of muscles. Yet feedback from the limb reaches both the α motor neurons and the spinal interneurons. These feedback pathways are indicated in Figure 11-2. How might these feedback pathways affect the deployment of muscle synergies?

When the interneurons in the spinal cord are electrically stimulated, they are presumably pinned at a high firing rate. For example, in the simplified diagram in Figure 11-2, suppose that interneuron 1 is electrically stimulated and thereby pinned at a high firing rate. Any feedback pathway that normally modulates the activity of interneuron 1 therefore becomes nonoperational. Within the circuitry controlled by interneuron 1, only feedback from the periphery to the α motor neurons remains operational. For example, stretch receptors in the muscles project directly to the α motor neurons, modulating their firing rate. By placing the limb in different positions, different patterns of stretch feedback are generated, and therefore the deployment of muscle synergy 1 could potentially be altered.

To examine the role of feedback in the generation of muscle synergies, Giszter et al. (1993) and Loeb et al. (1993) electrically stimulated the spinal cord of the frog to activate a muscle synergy, first with the sensory nerves from the leg intact and then with the sensory nerves cut. Removing the somatosensory feedback had a pronounced effect on some aspects of the evoked movement.

First, before the sensory nerves were cut, moving the leg to different positions changed the evoked synergy. Stimulation of a particular spinal site did not always evoke the same fixed pattern of muscle activity. The activity of each muscle depended to at least some degree on the position of the limb. When the sensory nerves were cut, these variations in muscle activity disappeared. Stimulation of a site in the spinal cord evoked a fixed ratio of muscle activity across a set of muscles, independent of the position of the limb.

Second, cutting the sensory nerves greatly reduced the amount of muscle activity and therefore the amount of force that was evoked by stimulation. To assess the evoked force, the foot was placed in a range of different locations and the evoked force was measured at each location. After deafferentation, this force field maintained roughly the same shape, its vectors aiming toward the same region of the workspace, but the magnitude of the evoked forces was reduced.

These two effects of deafferentation are consistent with a specific mechanism. First, the activation of a propriospinal neuron provides an instruction signal to the α motor neurons. Second, the instruction signal does not itself normally translate directly into muscle activity. Instead, it adjusts the continuously operating feedback loop between the α motor neurons and the muscle stretch receptors. As in the λ model, it is this feedback loop that determines the activity level of the α motor neurons and therefore of the muscles. If, however, the feedback from the limb is removed, then this circuitry is broken. All that is left is the instruction set. Normally the instruction set does not by itself determine the activity of the α motor neurons. Yet it does provide excitation to the α motor neurons, and it does so in a specific ratio consistent with the desired final position of the limb. Thus activation of the instruction set itself, in the absence of feedback to the α motor neurons, produces a kind of pale shadow of the normal movement, at low force, and without the appropriate position dependency of the muscle activity. In this view, muscle synergies combine elements of feed-forward and feedback control. The divergent mapping from a propriospinal neuron to a set of muscles provides a feed-forward signal. That feed-forward signal is adjusted continuously by feedback to the α motor neurons.

In this interpretation, however, the feedback loop between the α motor neurons and the muscle stretch receptors is only the most peripheral feedback loop of many. Artificial stimulation of the interneurons essentially removes all feedback loops except this most peripheral one. Operating by itself, its contribution is relatively weak. For this reason, experimenters may be tempted to dismiss feedback as a minor influence on the motor pathways.

BEYOND POSTURAL CONTROLLERS

The λ model and the model of muscle synergies are postural controllers. In the λ model, the system receives continuous feedback about muscle length. Muscle length is effectively a measure of joint angle. The feedback controller in the λ model therefore explicitly controls joint angle, seeking and defending a goal angle instructed by a higher-order signal. In the model of muscle synergies, a feed-forward signal provides excitation to a set of muscles in a fixed ratio. The effect of this fixed ratio of activity is a movement of the joints to an equilibrium position. The combination of the λ model and the muscle synergy model, discussed in the previous section, is therefore naturally also a postural controller.

In a postural controller, the fundamental building blocks of movement are postures. Other movement variables must be controlled by proxy, by creatively combining postures. For example, a curved trajectory of the hand through space might be controlled by tugging the hand toward one stored posture and then, partway through the trajectory, beginning to tug it toward another stored posture. Limb speed and force could be controlled by increasing or decreasing the gain on the neural signal that evokes the posture. A greater gain on the signal will make the limb move with greater acceleration toward the instructed posture. The position of the hand in space could be controlled by specifying the posture of the arm, so that the sum of the joint angles results in the desired hand position.

Yet the actual cortical-spinal-muscle system might not be a postural controller. The system does, of course, receive feedback about muscle length and therefore can control joint angle. But it also receives feedback about many other variables and can control these other variables as well. Sensors in the periphery supply information about the speed of muscle stretch, the force applied by the limb, and the pressure or pain on the skin as the limb presses against an object. Visual information about the limb also reaches the motor cortex and therefore can influence the spinal cord.

In addition to these direct sensory sources of feedback, the brain contains a higher-order representation of the structure, configuration, and movement of the body. This representation is sometimes termed the "body schema," or the internal model of the body (for review, see Graziano and Botvinick, 2002). The body schema is a highly integrative, multisensory representation. It brings together relevant information sources, including somatosensory information, visual information, motor control signals, and estimates of the shape and hinged structure of the body. These information sources are combined to generate an up-to-date and even a predictive model of the body as it moves in real time. The body schema provides an intelligent estimate of limb configuration. Many attributes of a body schema have been described in the parietal area 5, a cortical region that is mutually connected to the motor cortex, and in the neuronal properties within motor cortex itself. Wolpert and colleagues have elaborated on this hypothesis of an internal model of the body that is based on a Baysian summation of many sources of input and that is used to guide movement or predict the consequences of movement (e.g., Vercher et al., 2003; Wolpert et al., 1995). Ideally a feedback controller would use the information from the body schema and not rely solely on information from direct sensory feedback. Sensory feedback is delayed by a conduction latency, whereas the body schema can provide estimates that are current or even predictive. Moreover, a sensory signal, or any other sole source of information, is likely to be noisy and approximate. The job of the body schema is to generate superior quality information about the state of the periphery by intelligently combining all available sources of information. This information about the changing state of the body is certainly available to motor cortex and therefore to the spinal cord. It may be that some of the projections from the motor cortex to the spinal cord do not directly control movement but instead supply higher-order feedback about the body and limbs to supplement the somatosensory feedback that reaches the spinal cord directly.

With this rich infusion of feedback into every level of the motor system, there is no reason to hypothesize that the system is fundamentally a controller of posture. It is capable of controlling any variable for which it receives feedback. Because posture is often of great behavioral importance, it may be emphasized within the control system. Yet other movement variables can also be controlled independently of posture.

For example, consider pointing to a visual target. The hand must be brought to a specific end point in space. A postural controller might control hand position by specifying the joint angles of the arm. It might pick one set of joint angles that corresponds to the desired hand position and instruct the arm to acquire that set of joint angles. Yet in actual pointing, the angles of joints can vary considerably from trial to trial, whereas the position of the hand is quite reliable (Liu and Todorov, 2007; Todorov and Jordan, 2002). Clearly the system is controlling hand position more directly and not by proxy through posture. In this case, the system can use visual feedback about hand position to guide the movement of the hand. Todorov and Jordan (2002) argue that the motor system uses optimal feedback control. Feedback about task-relevant variables guides the movement, whereas other variables that are not directly task relevant are not as precisely controlled.

IS MOVEMENT CONTROL POSSIBLE WITHOUT FEEDBACK?

The previous section emphasized the role of feedback in the control of movement. Yet a surprising amount of motor control of the arm can be achieved without any somatosensory feedback. Taub et al. (1973; Taub et al., 1975) cut the sensory nerves from the arms of infant monkeys and observed them as they developed and learned to move. The deafferented monkeys were delayed in many motor skills by several weeks but learned to reach, climb, run, and in general move much like an intact monkey. To further study this phenomenon, Polit and Bizzi (1979) studied simple, visually guided reaching movements in adult monkeys that had had the sensory nerves from one arm cut. Each monkey sat in a chair with the deafferented arm in a specialized holder such that only the elbow joint could be rotated, moving the hand along a horizontal arc. The monkey's task was to aim its hand toward a set of target lights. As each light was illuminated, the monkey was required to point its hand at that light to receive a reward. Even when vision of the arm was blocked, the deafferented monkey was able to learn this task, although its pointing was much less accurate than the pointing of a normal monkey. These results demonstrate that some degree of spatial control of movement is possible without any somatosensory or visual feedback.

These results were taken as evidence for a feed-forward model. The controller, in that interpretation, is not a servo mechanism. It does not require online, continuous adjustment on the basis of feedback.

In an alternative interpretation, however, the control mechanism is indeed a feedback one, and feedback about arm position was available to the deafferented monkeys. Somatosensory feedback from the limb was removed. Direct visual feedback was also removed. However, the high-level representation of the body, the body schema or the internal model of the body, remained. Even in a monkey with no somatosensory or visual input from the arm, some sources of information remain to inform the body schema. Motor commands, when copied to the body schema, are informative about the likely position and movement of the limb. Sensation in the shoulder or in the upper torso may be informative. If the monkey sees the limb in the experimental apparatus before the limb is occluded, it will learn at least the general limb configuration. These sources of information may enable the calculation engine of the body schema to estimate the changing limb configuration during movement. This information, supplied by the body schema, could be used for online corrections by a servo or feedback controller.

The results of the deafferentation studies, therefore, do not necessarily argue against a feedback controller. The most general type of feedback controller would survive even as drastic a manipulation as a somatosensory deafferentation of the arm and a visual occlusion of the arm.

POSSIBLE EFFECTS OF CORTICAL STIMULATION

A simplified, schematic version of the cortical-spinal-muscle system is shown in Figure 11-3. In this scheme, neurons in motor cortex connect laterally to each other, with greater likelihood of connectivity among nearby neurons in a Gaussian neighborhood function. The cortical neurons also project divergently to interneurons in the spinal cord, and to α motor neurons in the spinal cord.

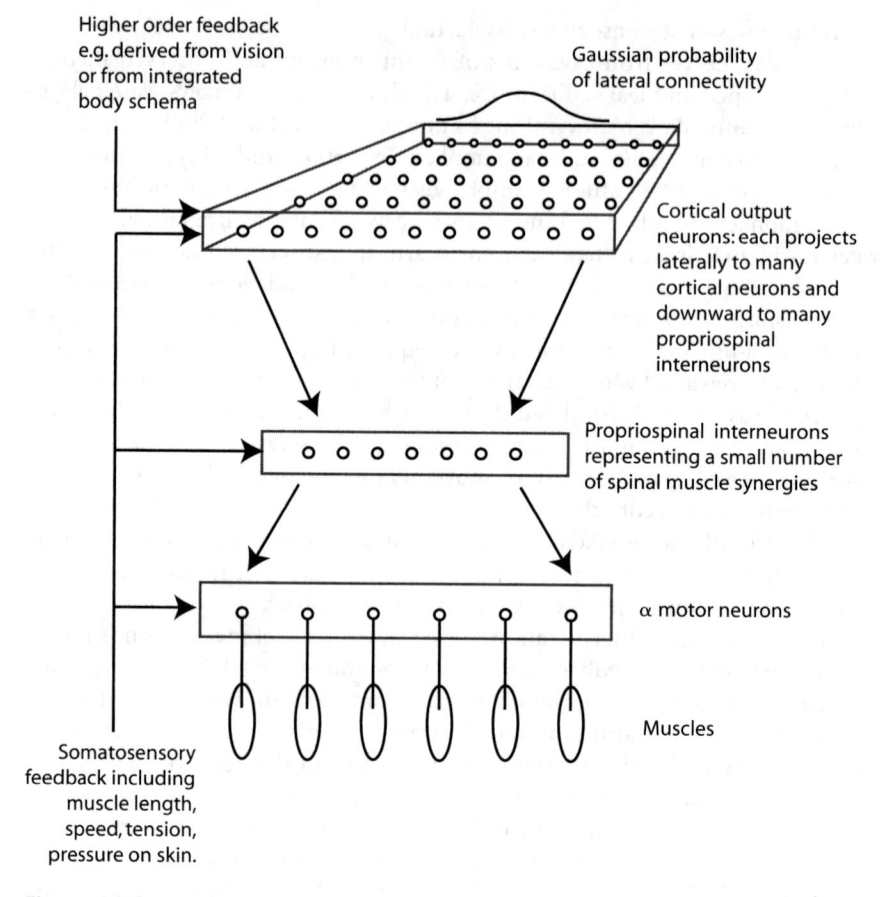

Figure 11-3 Simplified schematic of the cortical-spinal-muscle system.

The spinal interneurons project divergently to α motor neurons. Somatosensory feedback is sent directly to all of these neural elements including to the α motor neurons, spinal interneurons, and cortical neurons. Visual feedback reaches the motor cortex neurons. Higher-order information about the state of the body, including predictive information about the future state of the body, reaches the motor cortex. This simplified schematic captures some of the basic features of the cortical-spinal-muscle system.

Consider the effect of electrically stimulating a site in the motor cortex of this simplified model. If a single neuron is stimulated, it causes a branching recruitment of circuitry through the cortex and spinal cord, ultimately resulting in a set of muscle activations. Because neighboring neurons in cortex have a greater probability of interconnectivity, neighboring neurons tend to have an overlapping effect on the circuitry. Their functions are similar. Because of this similarity of neighboring neurons, stimulating a local cluster of cortical neurons (as in our experiments) should produce a coherent effect that is a blurring or averaging of the functions of the directly stimulated neurons.

In terms of the feed-forward effect, stimulation of a cortical site will recruit a branching tree of circuitry. Following the muscle synergy model, this branching tree of circuitry represents a useful linking of muscles, a relative distribution of activity across a set of muscles that is common in the repertoire of the animal and therefore useful to have stored within the circuitry. In this manner, a rough approximation of a normal movement can be evoked. For example, in our experiments, stimulation of a "reaching" cortical site might evoke activity in a set of muscles spanning the hand, arm, shoulder, and torso. The activation of such a large ensemble of muscles by one site in cortex is easily understandable given the divergence in the connectivity.

Yet the movement evoked by stimulation of a cortical site, in this model, will be more complex than solely a feed-forward activation of a set of muscles at a fixed ratio. The feedback pathways in the system will also affect the resultant movement. The neurons that are directly stimulated are presumably pinned at a high, steady rate of firing, thus nullifying any feedback signals that might normally modulate those neurons. In this sense, part of the feedback circuitry is disabled by the stimulation. Feedback is still free to modulate the activity of other cortical neurons outside the directly stimulated site, the activity of spinal interneurons, and the activity of α motor neurons. The many pathways through the network that connect the directly stimulated neurons to the muscles are all subject to modulation by feedback. The exact pattern of muscle activity that is evoked by stimulation of that site may therefore depend on the position of the limbs, the speed of joint rotation, the force applied by the limbs, pressure or pain feedback indicating the contact of a limb with external objects, and so on. In this manner, the motor output evoked by stimulation could be quite complex, transcending a fixed pattern of muscle output, resembling a complex, feedback-dependent piece of behavior. Such effects of feedback might explain our results such as the compensation of the arm for a weight fixed to the wrist, or the differing patterns of muscle activity depending on the starting position of the limb.

The stimulation-evoked movement would be dissimilar to a normal movement in several respects. First, as noted above, stimulation of a cluster of local neurons would result in a blurred, additive version of the effects of each individual neuron. For example, in our stimulation studies, a hand-to-mouth movement may be a blurred mean of a set of complex and varied hand-and-mouth-interactions. Second, the feedback circuitry would be partially crippled because the directly activated neurons become immune to modulation by feedback. Third, the directly activated neurons would not have a normal, complex temporal profile of activity. Instead their activity would rise rapidly to a plateau and remain fixed at that plateau during the period of stimulation. This lack of temporal modulation of the driving signal is almost certainly the reason why the muscle output in our experiments tended to follow a relatively squared plateau pattern.

The overall effect of cortical stimulation in this model, therefore, would be to send signal through a prelearned set of connections. The activated connectivity would be partly a feed-forward pathway diverging to affect many muscles, and partly a set of feedback loops. The consequence would be the activation of a fragment of behavior that was common enough to be learned and stored within the circuitry. The usefulness of storing that common movement fragment in the circuitry is that it can be used in combination with other stored movement fragments as part of the basis set for the general control of movement. The electrically evoked movement, however, would be an approximation to a normal movement. It would be a blurred average of many stored movements, and it would lack some of the nuance of feedback and of temporal patterning.

In this manner, by considering the known circuitry of the motor cortex and spinal cord, it is possible to arrive at a basic intuitive understanding of the processes set in motion by electrical stimulation, why the stimulation might evoke such complex movements, and why those movements resemble natural movements in some ways and not in others. The complexity of the movements evoked by stimulation of motor cortex is not mysterious. It is surprising only in the context of a traditional and clearly incorrect view of the system as a simple set of feed-forward cables.

Chapter 12

Social Implications of Motor Control

INTRODUCTION

The central theme of this book is that the primate motor system does not merely control muscle contractions but coordinates meaningful actions within the normal behavioral repertoire. The actions discussed throughout this book, such as hand-to-mouth actions, defensive actions, and reaching to grasp, are basic means of interacting with the inanimate environment. Does social behavior also partly depend on a set of complex, useful actions built into the motor machinery? One does not normally think of social skill as a function of the motor system. Yet many fundamental features of human social gesture and expression, such as smiling, laughing, crying, or making hand gestures during speech, may have evolved from actions long built into the primate motor system, as Darwin first suggested (1873). The following sections speculate on several points of contact between human social behavior and the control of basic motor repertoire. The chapter ends with a discussion of the possible relationship between malfunctions of the motor system and autism.

RELATIONSHIP BETWEEN DEFENSIVE REACTIONS AND SOCIAL DISPLAYS

An action mode of particular importance to survival is the defense of the body surface. We studied these reactions in fascicularis monkeys in particular quantitative detail (Cooke and Graziano, 2003, 2004a, b; Cooke et al., 2003). This class of behavior includes the ducking, squinting, and blocking movements made in response to stimuli that loom toward or touch the body. As we quantified these movements in monkeys we could not help but notice a formal similarity between them and many displays common in human social interaction. The general hypothesis that defensive reactions gave rise to many social displays was first elaborated by Andrew (1962). The following sections revisit this connection in light of our recent data.

Why would defensive reactions in particular have evolved into social displays? Defensive reactions have three specific properties that might make them especially likely to be adapted into social displays. First, the internal state of an animal affects the likelihood and magnitude of its defensive reactions. For example, physiological stress, recent episodes of startle, and perceived likelihood of attack will profoundly affect defensive reactions. Second, defensive reactions are easily visible and therefore broadcast this information about the inner state of an animal. Third, defensive reactions cannot be safely suppressed because they are necessary for survival. For these reasons, defensive behaviors provide a potential window into the inner state of an animal. When an animal makes a defensive reaction, it necessarily broadcasts information about itself that can potentially be exploited by conspecifics. This broadcasting of information about the inner state to conspecifics, and especially the manner in which that information modifies the behavior of conspecifics, presumably sets up evolutionary pressures. Two kinds of adaptations are expected. First, animals might evolve mechanisms for detecting and taking advantage of the information provided by the defensive reactions of others. Second, animals might evolve mechanisms for modifying their defensive reactions to influence the behavior of others. It has been pointed out that the adaptive value of social signals is not their information content by itself, but rather their impact on the behavior of the receiver (Dawkins and Krebs, 1978). In this way defensive movements may have evolved into a complex set of social signals.

Summary of Defensive Reactions

Two general kinds of defensive reactions can be distinguished. The first kind, the startle reflex, is fast, probably subcortically mediated, and generally places the body in a protective posture that is not sensitive to the spatial location or trajectory of the threat (Landis and Hunt, 1939; Yeomans et al., 2002). The startle reflex is by hypothesis such a rapid response, beginning within 10 ms of the sensory stimulus, that the neuronal machinery has no time for complex spatial computations. For example, startle is bilaterally symmetric regardless of the location of the stimulus.

A second class of defensive reaction is the spatially specific reaction that begins about 30 ms to 50 ms after the stimulus onset (e.g., Cooke and Graziano, 2003; Graziano and Cooke, 2006; Landis and Hunt, 1939; Schiff, 1965; Schiff et al., 1962; Strauss, 1929). This more complex reaction is probably cortically mediated. In many ways it resembles the startle reflex. All the same components can be present. However, the components are often directed toward the location or trajectory of the threat.

In humans and macaque monkeys, defensive reactions include a set of basic components (Cooke and Graziano, 2003; Davis, 1984; Landis and Hunt, 1939; Yeomans et al., 2002). These components may or may not be present in every case depending on the magnitude of the reaction or the specific nature of the stimulus. The components include

1. A squinting of the musculature around the eyes (the orbicularis muscle) often accompanied by a closure of the eyelids, presumably to protect the eyes. Tear production can also occur on a longer delay, depending on the nature of the stimulus.
2. A lifting of the upper lip and drawing backward of the corners of the upper lip, probably actuated by muscles including the nasolabialis and the zygomaticus. This facial movement results in the cheeks bunching or wrinkling upward toward the eyes and therefore may contribute to

12. Social Implications of Motor Control

the protection of the eyes. In some cases the jaw opens partly, or the lower lip pulls down exposing the lower teeth, but these components are inconsistent compared to the lifting of the upper lip.
3. A withdrawal of the eyeball into the socket by a small amount, caused by the cocontraction of extraocular muscles. A side effect of this ocular withdrawal is a centering of the eye in the orbit. The presumed function of the ocular withdrawal is protection of the eye.
4. A flattening of the external ear against the side of the head, seen consistently in macaques and occasionally in humans, presumably to protect the external ear.
5. A downward and forward ducking of the head accompanied by a lifting of the shoulders, thought to protect the neck, a body part particularly vulnerable to predation. In a poststartle reaction, the head may also turn away from the direction of the threat.
6. A forward curvature of the torso accompanied by a flexion of the hips and knees, thought to reduce the height of the body and therefore reduce its vulnerability as a predatory target.
7. Blocking movements of the forelimb to protect vulnerable parts of the body. In a startle reaction the most common movement is a drawing of the forelimbs forward and across the front of the torso, thought to help protect the vulnerable abdominal area. In a poststartle defensive reaction, the forelimbs may move rapidly up to protect the face or to block an impact on some other part of the body. A threat to the hand may cause the arm to retract rapidly toward the body or move behind the back.
8. A sharp exhalation, sometimes voiced, possibly a side effect of the rapid hunching of the body.

Defensive Reactions and Smile

It has been suggested that a human smile might be homologous to the "fear grimace" or "submissive grimace" seen in nonhuman primates (Andrew, 1962; van Hooff, 1972; Preuschoft, 1992). Because the term *grimace* is ambiguous, the more precise description of "silent bared teeth display" is often used (van Hooff, 1972; Preuschoft, 1992). This description, though more informative than "grimace," nonetheless focuses on the baring of the teeth and underemphasizes a large set of components that accompany the behavior. This submissive display in a fascicularis monkey closely resembles the defensive face-protecting reaction. In a submissive display the skin around the eyes becomes puckered in a partial squint, the upper lip pulls up and back at the corners thus causing the cheeks to bunch or wrinkle upward toward the eyes, the ears flatten back against the head, the head ducks down, and the shoulders shrug. Sometimes the hands are pulled across the front of the torso. In these respects the submissive action resembles the defensive action.

The primary difference between the submissive display of a fascicularis monkey and the normal defensive reaction is that, in a submissive display, the eyes remain wide open. This modification of the original defensive behavior has an obvious adaptive value. Closing the eyes when confronted by a potentially aggressive dominant conspecific, and therefore shutting off visual information about the potential threat, would presumably be maladaptive.

What might be the evolutionary path from a defense of the face against impact to a social signal? If a large and aggressive animal A is looming toward a small animal B, it is presumably adaptive for B to enact a defensive posture, protecting itself against possible injury. This defensive posture however does more than physically protect B. As a side effect it also broadcasts information about the internal state of B. This broadcasting of information has two evolutionary consequences.

First, it is adaptive for animal A to perceive and act on that information. Animal A could potentially deduce that B is fearful, likely to withdraw, and therefore not a threat. By exploiting the information gained, A could better navigate its social world, avoiding costly fights and gaining useful feedback about its own social status.

Second, it is adaptive for animal B to use its defensive behavior to exploit the reaction of A. By making a defensive gesture, B communicates a nonthreatening status to A and thereby avoids attack.

By hypothesis, these influences resulted in the evolution of the social smile. The process, however, was not merely the evolution of the smile, but the coevolution of the social reaction to a smile and the social production of a smile. Both are essential, interacting parts of the same story.

Probably the human smile itself has begun to diverge into several overlapping behaviors used differently in different contexts. In some social contexts, a smile is limited mainly to the mouth. In contrast, the squinting of the eyes, the bunching upward of the cheeks toward the eyes, and the lifting of the upper lip especially at the corners of the mouth, are all components of what has been termed a "Duchenne human smile" (Ekman et al., 1990), a smile that encompasses the musculature of the upper face and is not limited to the mouth. An intense smile can cause the eyes to nearly close, increasing the resemblance to a facial defensive reaction.

Defensive Reactions and Laughter

In a brilliant realization, Van Hooff (1962, 1972) first suggested that laughter in humans is homologous to a facial display seen in other primates, the relaxed, open-mouthed play display (see also Preuschoft, 1992). This display is speculated to have originated from play fighting in which the mouth is opened to bite gently and nonaggressively.

Consider the nature of play fighting. It involves two simultaneous, highly integrated processes: attack and defense. Attack during play fighting involves penetrating the defended regions of personal space of the other animal to touch or play bite vulnerable parts of the body. Defense during play fighting involves all the normal defensive reactions listed above that protect regions of the body and maintain a margin of safety. Each animal in the play fight must perform both of these processes simultaneously. In the hypothesis presented here, human laughter is not simply a ritualized version of a play bite, but instead a ritualized modification of the entire package, the play bite and the play defense. Indeed, the physical similarity between human laughter and a normal defensive reaction is especially close.

Consider the phenomenon of tickling in humans. It is a version of play fighting. The act of tickling is a sensory intrusion into vulnerable and normally protected regions of personal space. The intense laughter and fast protective redeployment of the limbs that occurs involuntarily in response to tickling resembles an exaggerated version of a normal defensive reaction. The components include a contraction of musculature around the eye and sometimes eye closure, tear production, a raising of the upper lip accompanied by a bunching of the cheeks upward toward the eyes, a ducking downward of the head and a shrugging upward of the shoulders, a hunching or forward curving of the torso, and a pulling of the arms inward across the vulnerable abdomen. The classical "ha ha ha" of laughter is a repetition of the sharp exhalation that occurs during a defensive reaction. Point by point, tickle-evoked laughter closely resembles the defensive reactions normally used to protect the body when it is threatened by intruding stimuli. The primary difference between tickle-evoked laughter and a normal defensive reaction is that the mouth is open in laughter, consistent with an origin in play fighting in which defensive reactions and play biting are integrated.

How might tickle-evoked laughter have evolved from a defensive reaction? Consider a play fight between animal A and B, in which B succeeds in touching, biting at, or scratching at a vulnerable body part of A. A produces a strong, normal defensive reaction. By producing a defensive reaction, and by the strength of the defensive reaction, A necessarily broadcasts that B has achieved a goal. B has succeeded in penetrating the defenses of A and has won a point in the play fight. The normal defensive reaction inevitably provides information that can potentially alter the behavior of B. Because of this potential for information transfer, evolutionary pressures are set up.

The receiver gains an adaptive advantage by exploiting the signal. Animal B can use the signal to advantage as positive feedback while training its fighting ability. The defensive action of A, like a "touché" signal, indicates that B has made a useful or successful move in the fight, and B's ability to attack and fight can be honed by exploiting this positive feedback. It is to B's advantage, therefore, to treat the signal as a social reward. The signal also indicates that B, having won the point, can stop its immediate play attack on that body location. Any further intrusion into personal space in that direction might injure the play partner and derail the play relationship. Because the play relationship is useful to B, it is adaptive for B to pull back on receiving the signal from A.

As a consequence of its ability to alter the behavior of the receiver, the signal also has adaptive value to the sender. It protects the sender from injury because it stops the receiver from pressing the play attack too far. It is to the sender's advantage to enhance the signal, ensuring personal safety. By enhancing the signal and making it more salient to the receiver, the sender exploits the receiver's exploitation of the sender's defensive reactions. The signal is shaped by these complex and recursive evolutionary pressures. On the sender's side, the signal becomes exaggerated to enhance its salience. On the receiver's side, it comes to be treated as socially rewarding feedback. In this manner it evolves into the exaggerated and ritualized interaction of tickle-evoked laughter in humans.

It is not difficult to imagine how tickle-evoked laughter could have given rise to the many other forms of human social laughter. In the hypothesis above, tickle-evoked laughter is a touché signal indicating that the tickler has achieved a success in interacting with the ticklee. The ticklee acknowledges the momentary success of the tickler. The signal is positive feedback to the tickler. Once such a signal has become established in the species, a complex and powerful dynamic is created. Each person has control of a social reward that he or she can dispense to other people. The reward is laughter. By selectively dispensing the reward, a person has some ability to shape the social behavior of others. For example, laughter at a joke or clever comment arguably rewards the speaker's mental prowess. Likewise, failure to laugh at a joke is a withholding of reward that can also shape the behavior of the speaker.

It is worth noting that a joke that hits the funny bone, so to speak, can evoke prolonged involuntary laughter. The laughter is far more than an open-mouthed facial display or a vocalization. It includes a contraction of musculature around the eye and sometimes eye closure, a raising of the upper lip causing a bunching of the cheeks upward toward the eyes, sometimes a ducking downward of the head and a shrugging upward of the shoulders, a forward curving of the torso, a pulling of the arms inward across the abdomen, tear production, and a repeated sharp exhalation. It has all the components of tickle-evoked laughter, and all the components of a normal defensive reaction. Laughter at a killer joke is, in this sense, a prolonged caricature of a defensive reaction.

It has been noted that in some instances human laughter includes a throwing back of the head, aiming the mouth upward (von Hooff, 1972). In contrast, a normal defensive reaction and some instances of human laughter involve the head ducking forward and downward. If one inspects video pans of audiences at stand-up comedy acts, one sees instances of both laughter patterns. One possibility is that human laughter has evolved into a range of related behaviors that are not identical. For example, it may be that tilting back the head is a modification of the original defensive-like pattern and serves to project the sound of the laughter, enhancing its signal strength. By this hypothesis, loud projective laughter in a group should be systematically associated with a back-tilted head, and quiet, giggling laughter between two individuals should be associated with a downward ducking of the head and shrugging of the shoulders.

The point of these far-flung speculations is not to present a specific theory of this or that kind of laughter, a task that would require much more data. The fundamental point here is that normal defensive behavior could plausibly have served as the origin of tickle-evoked laughter because tickling is an invasion of normally defended personal space, and because tickle-evoked laughter matches point-for-point the characteristics of a defensive reaction. Tickle-evoked laughter could then have diverged into the many social uses of laughter apparent in humans. A secondary point is that there is probably no single type of laughter or single explanation that can cover all instances of laughter. Rather it is more plausible to postulate a social display that is in a continual process of diversifying to perform many social functions.

Defensive Reactions and Crying

The act of crying, more than any other human social behavior, looks like a simulation of a defensive reaction. The behavior under discussion here is not a distress call such as many animals including humans make, or the wailing of a human infant that presumably falls into the category of a distress call. Instead the behavior under discussion is a squinting of the eyes, an excretion of tears, a lifting of the upper lip that results in an upward bunching of the cheeks toward the eyes, a ducking of the head, a shrugging of the shoulders, a forward curving of the torso, a flexion of the hips and knees, a pulling of the arms across the torso or upward over the face, and a sharp vocal exhalation. These components point-for-point resemble or are an exaggeration of an extreme defensive reaction.

It is interesting that the components of crying resemble the components of extreme laughter so closely that it is mainly social context that allows humans to distinguish the two states. This bizarre similarity between two social displays that have apparently opposite meanings was noted at least as far back as Homer, who, in *The Odyssey*, famously compared the laughter of a group of men to the crying they were about to do on being killed by Odysseus.

This similarity between crying and laughing suggests a further speculation. Crying could plausibly be an evolutionary modification of tickle-evoked laughter. In the hypothesis outlined in the previous section, tickle-evoked laughter evolved from play fighting in which a strong defensive reaction broadcasts that one animal has succeeded in penetrating the defenses of another animal and has contacted a vulnerable body part. The signal is not all-or-nothing. It is a graded signal, in which a stronger or more intense signal is evoked by a greater degree of violation of personal space. An extreme signal suggests such a violation of personal space that injury may have occurred. In a play fight, injury is clearly not an adaptive goal to either participant. An extreme defensive reaction could serve as a useful signal for the fight to stop and the winner to comfort the loser, to reestablish social amity. Because of its adaptive value, the signal is therefore put under evolutionary pressure. Two kinds of adaptations are expected. First, it would be adaptive for humans to evolve a strong and immediate response to the signal that includes comforting and providing help to the person whom one has accidentally injured. In this way, useful social amity is preserved. Second, it would be adaptive for humans to exploit the signal, using it to gain comfort and help even outside the context of a play fight and from individuals who were not responsible for the injury. The hypothesis proposed here, therefore, is not that crying is an extreme form of tickle-evoked laughter, but that it is an evolutionary modification of tickle-evoked laughter. The signal has taken on its own social role but physically retains its resemblance to its close ancestor, the tickle-evoked laughter.

Defensive Reactions and Personal Space

Hediger (1955) argued that escape was the most urgent survival requirement of any animal, trumping the more postponable functions of sex and eating. Through his observations of wild and captive animals, Hediger formulated the concept of a flight distance, now often called a "flight zone." In his formulation, escape is not a simple, stimulus-driven reflex. The sight of a predator is not enough to cause an animal to flee. Instead, the animal uses its active attention to its surroundings and its spatial cognition to construct a margin of safety around its body. When a threatening object enters this margin of safety or "flight zone," the animal escapes. Depending on the potential threat value of the intruding stimulus, the animal may simply adjust its position to reinstate the margin of safety, or may enter full flight. According to Hediger's observations, grazing animals have an especially large flight zone of tens of meters that can expand or contract depending on circumstances. A domesticated animal will in general have a much smaller flight zone. The concept of a flight zone has even been applied to the practical theory of cow herding (Smith, 1998).

Hediger's work on the flight zone led directly to the concept of personal space in humans. Many researchers noted that humans have an invisible bubble of protective space surrounding the body, generally larger around the head, extending farthest in the direction of sight (e.g., Dosey and Meisels, 1969; Hall, 1966; Horowitz et al., 1964; Sommer, 1959). When that personal space is violated, the person steps away to reinstate the margin of safety. Personal space, therefore, is the flight zone of a human with respect to other humans. The size of the personal space varies depending on context. A person who is placed in a potentially threatening context will have an expanded personal space; a person in friendly company will have a reduced personal space (Dosey and Meisels, 1969; Felipe and Sommer, 1966). In this view, personal space is fundamentally a protective space, a margin of safety. Intrusions into it trigger a defensive retraction. Personal space is therefore a particularly obvious example of the intersection between defensive behavior and social interaction.

Sex and Suppression of Defensive Mechanisms

Suppose animal A is motivated to mate with B. It is adaptive for A to reduce its personal space with respect to B. Obviously A must shrink its defenses enough to allow B into close proximity. Even more so, A must suppress its defensive reactions enough to allow direct and sustained skin contact with B. But the suppression of defensive mechanisms is not merely of mechanical utility. The change in behavior also contains information about A's level of interest in B. It is adaptive for B to perceive and interpret this change in the behavior of A. In this way B can learn that A is receptive and can even probe the specific level of receptivity by actively testing A's defensive mechanisms.

If B has such a reaction to A's behavior, then it is adaptive for A to exaggerate its behavioral signs of reduced defensive mechanisms to enhance B's detection of the signal. Because of these complex and interactive evolutionary pressures, specific signals of sexual receptivity evolve. In pair bonding, humans touch their mouths, a main offensive weapon, to each other's bodies with an emphasis on vulnerable portions such as the face, eyes, throat, hands, and abdomen. The more a body part is normally defended, the more it is voluntarily exposed to risk as a sign of receptivity to another, and the more it is touched and nipped to establish receptivity. The poses of women in fashion magazines tend to involve the head tilted and the neck exposed, as if offering to let the viewer's teeth onto the one body part most vulnerable to predation. (Vampires, with their penchant for biting necks, also traditionally have a sexual connotation.)

The same process might help explain sadomasochism, which is merely a slight exaggeration of the behavioral pattern of probing a partner's defensive reactions to establish receptivity, and of suppressing self-protection as a signal of receptivity.

Sympathetic Defensive Reactions

When an object looms toward or strikes the face, a characteristic set of defensive reactions occurs, as described above. A similar set of actions is commonly performed by a human observing injury to another human. If we see someone walk into a tree, or watch a victim having his teeth drilled by a dentist, we sympathetically produce a constellation of defensive actions including squinting, lifting the upper lip thus wrinkling the cheek upward, ducking and turning aside the head, shrugging the shoulders, and lifting a hand. Interestingly, the sympathetic defensive reaction can generalize from physical injury to any harmful event. When a friend tells us about a financial disaster, to express sympathy we in effect generate the facial actions that would normally protect us from collision.

The sympathetic defensive reaction may be an example of a mirror property previously described for grasp (Di Pellegrino et al., 1992; Gallese et al., 1996; Rizzolatti and Craighero, 2004). In monkeys and humans, some neurons in the motor system appear to combine sensory and motor properties. A mirror neuron that controls a specific type of grasp will also become active if the individual observes someone else performing the same type of grasp. This mirror property is hypothesized to play a role in the interpretation of the motor acts of other individuals. By simulating someone else's behavior with our own motor machinery in a subthreshold manner, we understand the behavior of the other individual.

The sympathetic defensive reaction suggests that at least in humans, mirror properties can be turned into an overt social signal. We not only use our motor machinery to understand the other individual, but also then explicitly signal that understanding to the other individual. The sympathetic defensive reaction communicates in effect, "I understand your distress, and to prove it I am generating the correct action myself for you to see."

Hypothesized Evolutionary Tree of Defensive Reactions and Social Offshoots

Normally an evolutionary cladogram shows a hypothesized branching of species. It may also be useful to draw a pseudocladogram showing the evolutionary branching of behavioral traits. This type of pseudocladogram must be taken with caution because behavioral traits could in principle influence each other or combine across branches in a manner impossible for a true cladogram of species. With these caveats in mind, Figure 12-1 helps to clarify the hypothetical relationships discussed above. The diagram shows defensive actions as a primitive trait that gave rise to a set of social behaviors. Social signals including laughing, crying, and sympathetic wincing share a family resemblance, though they are not physically identical. In the present hypothesis, the resemblance is due to their common origin in defensive reactions.

RELATIONSHIP BETWEEN COMMON ARM MOVEMENTS AND SOCIAL GESTURES

"Come Here" and "Hello"

In our observations of fascicularis monkeys described in Chapter 9, the reaching toward and acquiring of an object often included two components. One component was an extension of the hand away from the body with the wrist extended, the fingers opened, and the forearm pronated such that the palm faced outward. The second component was a scooping inward of the hand with a flexion of the wrist, curling of the fingers, and supination of the forearm such that the palm faced inward toward the body.

In humans the "scooping in" action has an obvious social meaning: "Come here," or "bring it here," as if the gesture represents a scooping of the desired object toward the body.

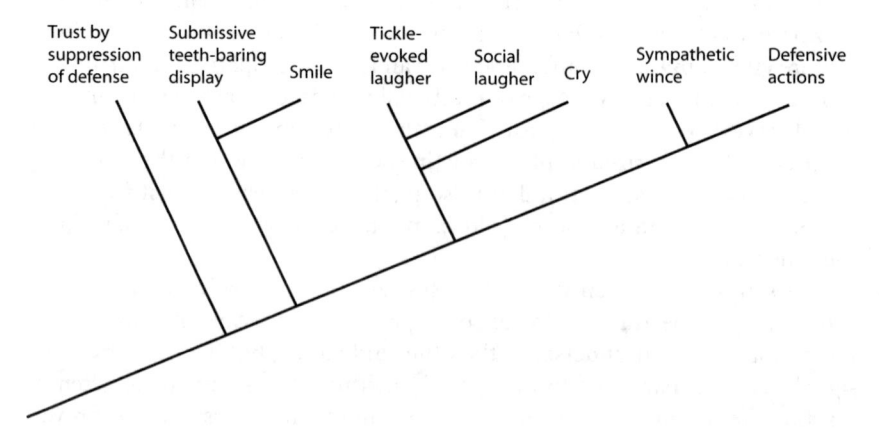

Figure 12-1 Hypothesized divergence of defensive actions into many social behaviors.

12. Social Implications of Motor Control

One could also argue that the standard hand gesture for *hello* or *goodbye* is a modification of reaching outward toward an object, with the hand opened, the wrist extended, and the forearm pronated such that the palm faces away from the body. When children learn to wave (at least the children in the daycare that my son attends), they extend the arm as if to reach and then make a typical infant grasping motion, bringing the four fingers in opposition to the palm. Infants before about the age of ten months do not typically have a precision grip involving the thumb. They reach out and grasp objects in the same manner in which they wave hello or goodbye.

A similar gesture of extending the arm with the hand open and facing outward is also used as a signal to stop approaching, stop walking, or stop talking (sometimes called the "talk to the hand" gesture). Why should a friendly gesture of greeting so closely resemble a gesture for stopping? One speculation is that the similarity is coincidental and that the two gestures originated from two different motor acts, the greeting originating from reaching out to grasp an object, the stop signal originating from the defensive blocking action.

Gesturing During Speech

As described in Chapter 9, in monkeys about 97% of grasp and manipulation occurs in central space near the chest or the mouth. This behavior is almost evenly divided into manipulation using the hand in central space in front of the chest, and manipulation between the hand and the mouth. The monkey, working on a toy or a piece of food, demonstrates a comical and incessant switching between these two regions of manual space. I could not help notice, while watching people in conversation, that normal hand motions during conversation follow almost the same statistics. Even if there is no object to be physically manipulated, the speaker will gesture with the hands in the central manipulatory space about 10 cm to 20 cm in front of the sternum, then lift one hand to the area of the mouth to tap the lip or chin, to stroke the beard, to hold a fist loosely against the mouth, to trace the line of the lips nervously with the forefinger, and so on. Up and down, the hand switches from central space in front of the sternum to the space at the mouth and back again. Punctuated throughout this behavior, occasionally the hand reaches out toward the other speaker with the characteristic opening of the grip, extension of the wrist, and pronation of the forearm that orients the palm away from the body. It is as if the same underlying motor machine, present in monkeys and evolved for certain practical uses, were on free run, generating the gestures during human speech.

When comparing good actors and bad actors, one sees a difference in the statistics of their gestures during speech. The good actor's gestures conform more or less to the statistics described above. The poor actor violates the natural statistics and makes gestures that are too often outside the common manipulatory regions near the chest and the mouth. The gestures are too large, the hand is too often distant from the body. In watching a student version of *Hamlet* a few years ago, I saw Hamlet hold a skull on his palm and grandly lift it to upper central space, in height just above his chin, at arm's length from the body, and proclaim, "Alas, poor Yorick!" This classic example of stagy acting had, at its root, an error in hand placement. The skull should have been held near the body at about sternum level, in the region of space in which the hand most often grasps and manipulates objects or holds them for inspection. In that case, the gesture would have seemed natural.

LINK BETWEEN AUTISM AND MOTOR CONTROL

Social behavior is partly a repertoire of species-specific actions that are deployed in specific circumstances. Some of these actions, such as allowing other people into normally defended personal space, making sympathetic winces, smiling, laughing, crying, and gesturing with arms and hands, are discussed above. The dependence of social interaction on a basic motor repertoire begs the question: what happens to social interaction when the machine that organizes and generates the motor repertoire is disturbed?

Autism Syndrome

Autism affects about 0.2% of the population. The incidence is three to five times higher for males than for females. The deficit is life long, though symptoms can improve over time with training. In the classical description (Kanner, 1943; Volkmar et al., 2005), autism is primarily a deficit in social and linguistic development. The syndrome is typically first diagnosed around the age of two to three years when infants begin to interact with peers in larger social settings, although parents often report noticing that something is subtly wrong before then. The symptoms are diverse and can vary from child to child, making the diagnosis difficult in some cases. Generally, however, three broad classes of symptoms have been described. The first class of symptoms includes a lack of speech, or slowing of speech development, or in some cases an apparent degradation of the speech ability that has already developed by the first few years. The second class of symptoms involves an apparent lack of interest in or an active avoidance of social interactions. For example, the child may avert its gaze to avoid looking at other people, sit in a corner, or pull away from physical contact. Normal social behavior never develops. As part of this social dysfunction, emotional disturbances such as socially inappropriate rage are also common. The third class of symptoms involves an obsessive focus on routine tasks, rigid order, and sameness. People with autism might spend a large proportion of time in tasks such as sorting or counting, or engaging in repetitive motor actions such as hand flapping and rocking.

Motor Deficits in Autism

The classical description of autism does not include any reference to a specific motor deficit. The abnormal movements are often attributed to psychological origins. In the classical view, hand flapping and rocking are part of the autistic

12. Social Implications of Motor Control

personality that seeks repetition and sameness. Yet when analyzed systematically, children with autism typically show abnormal motor control. Indeed, the motor disabilities are often glaringly obvious. In addition to repetitive movements such as hand flapping and rocking, the motor symptoms include abnormal postures of the arms and upper body and poor postural stability (Kohen-Raz et al., 1992; Minshew et al., 2004), abnormalities in the kinematics of gait during walking (Hallett et al., 1993; Vilenski et al., 1981), abnormally slow or delayed saccadic and smooth pursuit eye movements (Goldberg et al., 2002; Minshew et al., 1999; Takarae et al., 2004a, b), some deficits in reaching to grasp (Mari et al., 2003), some deficits in manipulation and grip strength (Hardan et al., 2003), and sometimes a verbal dyspraxia such as an inability to form phonemes correctly (Gernsbacher, 2004).

These motor deficits do not appear to be secondary symptoms caused by psychological factors such as a desire to withdraw, to socially repel other people, or to express aggression. Rather, the motor deficits can be observed early in life, long before the social and emotional symptoms are usually diagnosed, and in some cases might even be present at birth. Teitelbaum et al. (1998; Teitelbaum et al., 2004) collected and analyzed videos that parents had taken of their children during infancy. The children included some who developed normally and some who went on to develop autism or Asperger's syndrome (considered to be a mild form of autism). In analyzing these videos, Teitelbaum et al. found a set of motor deficits in the autism and Asperger group that were not present in the normal comparison group. For example, the infants in the autism group were unable to turn normally from their backs to their stomachs. Even those infants in the autism group who succeeded at righting themselves did so with abnormal postures of the limbs and rigidity of the torso. In another test, when held upright and then tilted laterally, unlike normally developing infants, the infants in the autism group did not counter-rotate their heads but maintained the head in rigid alignment with the axis of the torso. Even shortly after birth, some of the children demonstrated motor abnormalities in the shaping of the mouth. These results suggest that motor disability may be an early property of autism in at least some cases.

As pointed out by Teitelbaum et al. (1998), the literature abounds with anecdotes of children with autism who can climb to high places, stack blocks in impressively tall towers, or play the piano with astonishing technical accuracy. For this reason, perhaps, the association between autism and motor deficits has been slow to be recognized, with some practitioners claiming that individuals with autism have unusually good motor skills. The essential point is not whether these motor tasks can be accomplished, but how they are accomplished. There is a lack of the normal kinematic fluidity. The posture of the head, trunk, and arms is awkward. The repetitive movements are themselves deficits in the deployment of basic motor repertoire. The lack of eye contact, reduction of normal facial social displays, and slow development of communicative hand gestures such as pointing are all deficiencies in motor repertoire.

Leary and Hill (1996) suggested that the motor symptoms of autism are typically overinterpreted and incorrectly labeled as social symptoms. For example, a child with autism may have ataxia of facial muscles. In a social context, this symptom will be overinterpreted as a lack of appropriate emotion. Likewise, the child with autism might engage in hand flapping or other stereotypic movements that are overinterpreted as hostility or social withdrawal. In this manner the motor deficits may go unnoticed or unappreciated. Children with autism are sometimes also diagnosed with developmental motor disability, or "clumsy child syndrome." Yet even when children with autism are not "clumsy" and are therefore not diagnosed with developmental motor disability, they may still have profound motor impairments that are misattributed to cognitive, social, or emotional reasons.

Motor Hypothesis of Autism

The hypothesis suggested here, in its strongest form, is that social interaction is not solely a process of high-level cognition and intentional choices but also depends on a set of stereotyped actions built into the human motor system through millions of years of evolution. The use of gaze as a social signal, communicative pointing, and facial displays such as smiles, are all presumably disrupted by dysfunctions of the motor system. For this reason, motor dysfunction and social dysfunction often co-occur.

The hypothesis is not that autism results from an underlying weakness of the muscles, clumsiness, or lack of smooth or coordinated control of muscles. The hypothesis is not that the high-order cognitive and social deficits in autism can be reduced to or explained by low-order motor deficits. Instead, the hypothesis is that the motor system is an active participant in high- and low-order coordination, in controlling muscles and in organizing meaningful behavioral repertoire. Deficits of the motor system can therefore affect the organization of basic social behavior.

In the present hypothesis, damage to the central motor machinery for producing a smile does not merely remove the ability to make the appropriate muscle contractions but removes smile from the repertoire. It removes the natural concept of smile.

The hypothesis of a specific link between autism and the motor system is not new. The following sections summarize several other views of autism that also suggest a link to motor control.

Cerebellar Abnormalities

A large number of brain imaging studies have examined whether individuals with autism have a specific locus of brain damage. Although evidence exists for structural abnormalities in many cortical and subcortical areas, the impression that emerges from the very diverse literature is that cerebellar abnormalities are the most consistent, or at least the most consistently reported. The abnormalities of the cerebellum are typically most pronounced in the posterior vermis (Ahsgren et al., 2005; Courchesne et al., 1988; Courchesne, Saitoh, et al., 1994; Fatemi et al., 2002; Gaffney et al., 1987; Hashimoto et al., 1993; Kaufmann et al., 2003; Piven et al., 1992). In some reports, the cerebellum is smaller than normal, in other reports larger, perhaps corresponding to two subtypes that can be separately distinguished in the population (Courchesne, Saitoh, et al., 1994). Individuals with autism also have fewer and smaller cerebellar Purkinje cells (Bailey et al., 1998; Fatemi et al., 2002; Ritvo et al., 1986). Because the cerebellum is densely connected to the rest of the motor system, structural abnormalities in the cerebellum would presumably result in functional abnormalities throughout the entire motor control network, including cortical and subcortical components.

Mirror Neurons

It has been suggested that autism might be linked to a deficit in "mirror neurons" (Dapretto et al., 2006; Williams et al., 2001). Mirror neurons were first described in the ventral premotor cortex of monkeys. These neurons respond when the monkey performs a specific motor act such as grasping an object, and also when the monkey observes someone else performing the same motor act (Di Pellegrino et al., 1992; Gallese et al., 1996; Rizzolatti and Craighero, 2004). Similar mirror-neuron activity has been obtained in humans in imaging experiments (Iacoboni et al., 1999). The presence of mirror neurons suggests that the motor cortex machinery for producing actions may also be used to comprehend the actions of others. If the mirror neurons embedded in the motor system are used to understand the gestures, goals, and intents of others, then individuals with autism, who classically lack insight into the goals and motives of others, may lack normal mirror neurons. Individuals with autism are impaired at imitating the actions of others (Charman et al., 1997; Rogers et al., 2003; Stone et al., 1997) and appear to have a relative lack of mirror-neuron activity (Dapretto et al., 2006). Whether the mirror-neuron deficit caused the autism or the other way around, however, is not yet known.

The motor hypothesis of autism described above is in some ways the flip side of the mirror-neuron hypothesis of autism. The two views are complimentary. In the motor hypothesis, the motor system organizes social gesture and display, and therefore dysfunction of the motor system prevents the individual from producing normal social displays. In the mirror-neuron hypothesis, the motor system not only generates social signals, but also participates in comprehending the social signals of others. Thus dysfunction of the motor system leads also to a deficit in social perception.

Gaze Control

Cortical and subcortical eye movement areas were once thought only to trigger and guide saccades but are now known to feed back onto sensory areas and help guide the locus of visual attention (Moore and Fallah, 2004; Muller et al., 2005). Visual attention moves transiently and automatically to the location of a planned saccadic eye movement, just before the saccade is executed (Hoffmann and Subramaniam, 1995; Kowler et al, 1995; Rizzolatti et al., 1987;

Shepherd et al., 1986). Furthermore, direct electrical stimulation of eye movement areas, such as the FEF and the superior colliculus, actually redirects the locus of visual attention (Moore and Fallah, 2004; Muller et al., 2005). This stimulation of eye movement areas generates a signal that feeds back onto sensory areas and modulates neuronal activity (Moore and Armstrong, 2003). These results suggest that the motor machinery for moving the eyes is an integral part of the machinery for directing attention. This motor theory of attention is one example of the apparent inextricability of movement control and high-level cognition. Because individuals with autism are impaired on gaze control, they should be expected to have deficits in directing attention to new locations, such as to other people or to objects that are pointed out to them. This type of attentional deficit is common in autism (Courchesne, Townsend, et al., 1994; J. Townsend et al., 1996). It was traditionally attributed to psychological origins; the individual "doesn't want" to engage with another person, or "refuses" to be drawn out. An alternative explanation is that the gaze control mechanism is not functioning correctly, and thus the individual has trouble in rapidly and easily directing attention to targets.

A motor deficit in gaze control, in addition to causing a general attentional deficit, might be particularly disruptive for social development. Gaze is important in social communication (Argyle and Cooke, 1976; Macrae et al, 2002). Pointing with the eyes is used constantly during communication, and eye contact is used to communicate interest in another individual. Because saccade, smooth pursuit, and fixation control are impaired in individuals with autism (Dalton et al., 2005; Goldberg et al., 2002; Minshew et al., 1999; Takarae et al., 2004a, b), perhaps these children never have the chance to develop normal social interaction skills. In this view a motor dysfunction impairs one of the most basic forms of non-verbal communication.

Extreme Male Brain

Arguably the most plausible current explanation of autism is that it is an exaggeration of the normal traits of a masculinized brain (Baron-Cohen et al., 2005). Hormonal signals are known to shape brain development. Under the influence of testosterone, male brains develop in a specific manner. In the hypothesis of Baron-Cohen et al., autism occurs when the hormonally induced male differentiation of the brain becomes exaggerated. Consistent with this hypothesis, autism is three to five times more prevalent in males. Moreover, many traits of autism resemble exaggerations of male stereotypes, such as relative lack of social communicative ability and an obsession with systematizing. More recently it has been found that women with autism show symptoms typical of testosterone overexposure (Ingudomnukul et al., 2007). The motor hypothesis of autism is consistent with this overmasculanization theory of autism. It may be that one of the principle effects of androgens on brain development is to partially reorganize the motor system of males, equipping them for a different set of motor demands. By hypothesis, in autism the motor reorganization proceeds to a maladaptive extreme.

12. Social Implications of Motor Control

High Level Versus Low Level in Motor Control

The conundrum of motor deficits in autism lies in the apparent gap between low-level motor symptoms and high-level cognitive and social symptoms. What explanation can bridge the gap? The answer may be that motor control is not solely low level. The motor system, in the view outlined in this book, is an active participant in generating coordinated and useful behavior, organizing motor repertoire, contributing to spatial attention, and participating in the perception of the actions of others. It is unlikely that damage to the motor system is the sole cause of autism. Autism is diverse, and many brain areas may play a role to different degrees in different individuals. However, the cortical motor areas should probably be listed among the brain regions critical for normal social ability. Social behavior is above all a motor repertoire.

Literature Cited

Aflalo, T.N., and Graziano, M.S.A. (2006a). Partial tuning of motor cortex neurons to final posture in a free-moving paradigm. *Proc. Natl. Acad. Sci. 103*: 2909–2914.

Aflalo, T.N., and Graziano, M.S.A. (2006b). Possible origins of the complex topographic organization of motor cortex: reduction of a multidimensional space onto a 2-dimensional array. *J. Neurosci. 26*: 6288–6297.

Aflalo, T.N., and Graziano, M.S.A. (2007). Relationship between unconstrained arm movement and single neuron firing in the macaque motor cortex. *J. Neurosci. 27*: 2760–2780.

Ahsgren, I., Baldwin, I., Goetzinger-Falk, C., Erikson, A., Flodmark, O., and Gillberg, C. (2005). Ataxia, autism, and the cerebellum: a clinical study of 32 individuals with congenital ataxia. *Dev. Med. Child Neurol. 47*: 193–198.

Andrew, R.J. (1962). The origin and evolution of the calls and facial expressions of the primates. *Behaviour 20*: 1–107.

Argyle, M., and Cooke, M. (1976). *Gaze and mutual gaze*. Cambridge, UK: Cambridge University Press.

Asanuma, H. (1975). Recent developments in the study of the columnar arrangement of neurons within the motor cortex. *Physiol. Rev. 55*: 143–156.

Asanuma, H., and Arnold. A.P. (1975). Noxious effects of excessive currents used for intracortical microstimulation. *Brain Res. 96*: 103–107.

Asanuma, H., and Rosen, I. (1972). Topographical organization of cortical efferent zones projecting to distal forelimb muscles in the monkey. *Exp. Brain Res. 14*: 243–256.

Asanuma, H., and Sakata, H. (1967). Functional organization of a cortical efferent system examined with focal depth stimulation in cats. *J. Neurophysiol. 30*: 35–54.

Asanuma, H., and Ward, J.E. (1971). Pattern of contraction of distal forelimb muscles produced by intracortical stimulation in cats. *Brain Res. 27*: 97–109.

Asatrayan, D.G., and Feldman, A.G. (1965). Functional tuning of the nervous system with control of movements or maintenance of a steady posture: I. Mechanographic analysis of the work of the joint on execution of a postural task. *Biophysics 10*: 925–935.

Bailey, A., Luthert, P., Dean, A., Harding, B., Janota, I., Montgomery, M., Rutter, M., and Lantos, P. (1998). A clinicopathological study of autism. *Brain 121*: 889–905.

Baron-Cohen, S., Knickmeyer, R.C., and Belmonte, M.K. (2005). Sex differences in the brain: implications for explaining autism. *Science 310*: 819–823

Bates, J.F., and Goldman-Rakic, P.S. (1993). Prefrontal connections of medial motor areas in the rhesus monkey. *J. Comp. Neurol. 336*: 211–228.

Beevor, C. (1888). A further minute analysis by electrical stimulation of the so-called motor region of the cortex cerebri in the monkey (Macacus sinicus). *Phil. Trans. R. Soc. Lond. B 179*: 205–256.

Beevor, C., and Horsley, V. (1887). A minute analysis (experimental) of the various movements produced by stimulating in the monkey different regions of the cortical centre for the upper limb, as defined by Professor Ferrier. *Phil. Trans. R. Soc. Lond. B 178*: 153–167.

Beevor, C., and Horsley, V. (1890). An experimental investigation into the arrangement of excitable fibres of the internal capsule of the bonnet monkey (Macacus sinicus). *Phil. Trans. R. Soc. Lond. B 181*: 49–88.

Beisteiner, R., Windischberger, C., Lanzenberger, R., Edward, V., Cunnington, R., Erdler, M., Gartus, A., Streibl, B., Moser, E., and Deecke, L. (2001). Finger somatotopy in human motor cortex. *Neuroimage 13*: 1016–1026.

Bizzi, E., and Mussa-Ivaldi, F.A. (1989). Motor control. In: *Handbook of Neuropsychology*, Vol. 2. Boller, F., and Grafman, J. (Eds.), pp. 229–244. The Netherlands: Elsevier.

Bortoff, G.A., and Strick, P.L. (1993). Corticospinal terminations in two new-world primates: further evidence that corticomotoneuronal connections provide part of the neural substrate for manual dexterity. *J. Neurosci. 13*: 5105–5118.

Boussaoud, D. (1995). Primate premotor cortex: modulation of preparatory neuronal activity by gaze angle. *J. Neurophysiol. 73*: 886–890.

Brasted, P.J., and Wise, S.P. (2004). Comparison of learning-related neuronal activity in the dorsal premotor cortex and striatum. *Eur. J. Neurosci. 19*: 721–740.

Brecht, M., Schneider, M., Sakmann, B., and Margrie, T.W. (2004). Whisker movements evoked by stimulation of single pyramidal cells in rat motor cortex. *Nature 427*: 704–710.

Brinkman, C. (1981). Lesions in supplementary motor area interfere with a monkey's performance of a bimanual coordination task. *Neurosci. Lett. 27*: 267–270.

Broca, P. (1960). Remarks on the seat of the faculty of articulate language, followed by an observation of aphemia (Tr. G. von Bonin). In: *Some papers on the cerebral cortex*, von Bonin, G. (Ed.). Springfield, IL: Charles C Thomas, pp. 49–72. (Original work published 1861 in *Bulletin de la Societe Anatomique de Paris 6*: 330–357)

Brodmann, K. (1909). *Vergleichende Lokalisationslehre der grosshirnrinde* [Comparative localization in the cerebral hemispheres]. Leipzig, Germany: J. A. Barth.

Bruce, C.J., Goldberg, M.E., Bushnell, M.C., and Stanton, G.B. (1985). Primate frontal eye fields. II. Physiological and anatomical correlates of electrically evoked eye movements. *J. Neurophysiol. 54*: 714–734.

Bucy, P.C. (1933). Electrical excitability and cyto-architecture of the premotor cortex in monkeys. *Arch. Neurol. Psychiat. 30*: 1205–1225.

Bucy, P.C. (1935). A comparative cytoarchitectonic study of the motor and premotor areas in the primate. *J. Comp. Neurol. 62*: 293–331.

Caggiula, A.R., and Hoebel, B.G. (1966). "Copulation-reward site" in the posterior hypothalamus. *Science 153*: 1284–1285.

Caminiti, R., Johnson, P.B., and Urbano, A. (1990). Making arm movements within different parts of space: dynamic aspects in the primate motor cortex. *J. Neurosci. 10*: 2039–2058.

Campbell, A.W. (1905). *Histological studies on the localization of cerebral function*. Cambridge, UK: Cambridge University Press.

Chakrabarty, S., and Martin, J.H. (2000). Postnatal development of the motor representation in primary motor cortex. *J. Neurophysiol. 84*: 2582–2594.

Charman, T., Swettenham, J., Baron-Cohen, S., Cox, A., Baird, G., and Drew, A. (1997). Infants with autism: an investigation of empathy, pretend play, joint attention, and imitation. *Dev. Psychol. 33*: 781–789.

Literature Cited

Chen, L.L. (2006). Head movements evoked by electrical stimulation in the frontal eye field of the monkey: evidence for independent eye and head control. *J. Neurophysiol.* 95: 3528–3542.

Chen, L.L., and Walton, M.M. (2005). Head movement evoked by electrical stimulation in the supplementary eye field of the rhesus monkey. *J. Neurophysiol.* 94: 4502–4519.

Cheney, P.D., and Fetz, E.E. (1985). Comparable patterns of muscle facilitation evoked by individual corticomotoneuronal (CM) cells and by single intracortical microstimuli in primates: evidence for functional groups of CM cells. *J. Neurophysiol.* 53: 786–804.

Cheney, P.D., Fetz, E.E., and Palmer, S.S. (1985). Patterns of facilitation and suppression of antagonist forelimb muscles from motor cortex sites in the awake monkey. *J. Neurophysiol.* 53: 805–820.

Churchland, M.M., and Shenoy, K.V. (2007). Temporal complexity and heterogeneity of single-neuron activity in premotor and motor cortex. *J. Neurophysiol.* 97: 4235–4257.

Churchland, M.M., Yu, B.M., Ryu, S.I., Santhanam, G., and Shenoy, K.V. (2006). Neural variability in premotor cortex provides a signature of motor preparation. *J. Neurosci.* 26: 3697–3712.

Cisik, P., and Kalaska, J.F. (2005). Neural correlates of reaching decisions in dorsal premotor cortex: specification of multiple direction choices and final selection of action. *Neuron* 45: 801–814.

Classen, J., Liepert, J., Wise, S.P., Hallett, M., and Cohen, L.G. (1998). Rapid plasticity of human cortical movement representation induced by practice. *J. Neurophysiol.* 79: 1117–1123.

Cohen, J., Cohen, P., West, S.G., and Aiken, L.S. (2003). *Applied multiple regression/correlation analysis for the behavioral sciences*, 3rd ed. Mahwah, NJ: Lawrence Erlbaum Associates.

Colby, C.L., Duhamel, J.R., and Goldberg, M.E. (1993). Ventral intraparietal area of the macaque: anatomic location and visual response properties. *J. Neurophysiol.* 69: 902–914.

Cooke, D.F., and Graziano, M.S.A. (2003). Defensive movements evoked by air puff in monkeys. *J. Neurophysiol.* 90: 3317–3329.

Cooke, D.F., and Graziano, M.S.A. (2004a). Sensorimotor integration in the precentral gyrus: Polysensory neurons and defensive movements. *J. Neurophysiol.* 91: 1648–1660.

Cooke, D.F., and Graziano, M.S.A. (2004b). Super-flinchers and nerves of steel: Defensive movements altered by chemical manipulation of a cortical motor area. *Neuron* 43: 585–593.

Cooke, D.F., Taylor, C.S.R., Moore, T., and Graziano, M.S.A. (2003). Complex movements evoked by microstimulation of Area VIP. *Proc. Natl. Acad. Sci. USA* 100: 6163–6168.

Courchesne, E., Saitoh, O., Yeung-Courchesne, R., Press, G.A., Lincoln, A.J., Haas, R.H., and Schreibman, L. (1994). Abnormality of cerebellar vermian lobules VI and VII in patients with infantile autism: identification of hypoplastic and hyperplastic subgroups with MR imaging. *Am. J. Roentgenol.* 162: 123–130.

Courchesne, E., Townsend, J., Akshoomoff, N.A., Saitoh, O., Yeung-Courchesne, R., Lincoln, A.J., James, H.E., Haas, R.H., Schreibman, L., and Lau, L. (1994). Impairment in shifting attention in autistic and cerebellar patients. *Behav. Neurosci.* 108: 848–865.

Courchesne, E., Yeung-Courchesne, R., Press, G.A., Hesselink, J.R., and Jernigan, T.L. (1988). Hypoplasia of cerebellar vermal lobules VI and VII in autism. *N. Engl. J. Med. 318*: 1349–1354.

Cramer, N.P., and Keller, A. (2006). Cortical control of a whisking central pattern generator. *J. Neurophysiol. 96*: 209–217.

Crammond, D.J., and Kalaska, J.F. (1996). Differential relation of discharge in primary motor cortex and premotor cortex to movements versus actively maintained postures during a reaching task. *Exp. Brain Res. 108*: 45–61.

Dalton, K.M., Nacewicz, B.M., Johnstone, T., Schaefer, H.S., Gernsbacher, M.A., Goldsmith, H.H., Alexander, A.L., and Davidson, R.J. (2005). Gaze fixation and the neural circuitry of face processing in autism. *Nat. Neurosci. 8*: 519–526.

Dapretto, M., Davies, M.S., Pfeifer, J.H., Scott, A.A., Sigman, M., Bookheimer, S.Y., and Iacoboni, M (2006). Understanding emotions in others: mirror neuron dysfunction in children with autism spectrum disorders. *Nat. Neurosci. 9*: 28–30.

Darwin, C. (1873). *The expression of the emotions in man and animals.* New York: D. Appleton and Company.

D'Avella, A., Saltiel, P., and Bizzi, E. (2003). Combinations of muscle synergies in the construction of a natural motor behavior. *Nat. Neurosci. 6*: 300–308.

Davis, M. (1984). The mammalian startle response. In: *Neural mechanisms of startle behavior.* Eaton, R.C. (Ed.). New York: Plenum Press, pp. 287–351.

Dawkins, R., and Krebs, J.R. (1978). Animal signals: information or manipulation? In: *Behavioral ecology: An evolutionary approach.* Krebs, R., and Davies, N.B. (Eds.). Oxford, UK: Blackwell, pp. 282–309.

Dean, P., Redgrave, P., and Westby, G.W. (1989). Event or emergency? Two response systems in the mammalian superior colliculus. *Trends Neurosci. 12*: 137–147.

Dechent, P., and Frahm, J. (2003). Functional somatotopy of finger representations in human primary motor cortex. *Hum. Brain. Mapp. 18*: 272–283.

di Pellegrino, G., Fadiga, L., Fogassi, L., Gallese, V., and Rizzolatti, G. (1992). Understanding motor events: a neurophysiological study. *Exp. Brain Res. 91*: 176–180.

Donoghue, J.P., Leibovic, S., and Sanes, J.N. (1992). Organization of the forelimb area in squirrel monkey motor cortex: representation of digit, wrist, and elbow muscles. *Exp. Brain Res. 89*: 1–19.

Dosey, M.A., and Meisels, M. (1969). Personal space and self-protection. *J. Pers. Soc. Psychol. 11*: 93–97.

Douglas, R.J., and Martin, K.A. (1991). A functional microcircuit for cat visual cortex. *J. Physiol. 440*: 735–769.

Duhamel, J.R., Colby, C.L., and Goldberg, M.E. (1998). Ventral intraparietal area of the macaque: congruent visual and somatic response properties. *J. Neurophysiol. 79*: 126–136.

Dum, R.P., and Strick, P.L. (1991). The origin of corticospinal projections from the premotor areas in the frontal lobe. *J. Neurosci. 11*: 667–689.

Dum, R.P., and Strick, P.L. (2002). Motor areas in the frontal lobe of the primate. *Physiol. Behav. 77*: 677–682.

Dum, R.P., and Strick, P.L. (2005). Frontal lobe inputs to the digit representations of the motor areas on the lateral surface of the hemisphere. *J. Neurosci. 25*: 1375–1386.

Durbin, R., and Mitchison, G. (1990). A dimension reduction framework for understanding cortical maps. *Nature 343*: 644–647.

Ekman, P., Davidson, R.J., and Friesen, W.V. (1990). The Duchenne smile: emotional expression and brain physiology. II. *J. Pers. Soc. Psychol. 58*: 342–353.

Ethier, C., Brizzi, L., Darling, W.G., and Capaday, C. (2006). Linear summation of cat motor cortex outputs. *J. Neurosci. 26*: 5574–5581.

Evarts, E.V. (1968). Relation of pyramidal tract activity to force exerted during voluntary movement. *J. Neurophysiol. 31*: 14–27.

Fatemi, S.H., Halt, A.R., Realmuto, G., Earle, J., Kist, D.A., Thuras, P., and Merz, A. (2002). Purkinje cell size is reduced in cerebellum of patients with autism. *Cell. Mol. Neurobiol. 22*: 171–175.

Feldman, A.G. (1966). Functional tuning of the nervous system with control of movement or maintenance of a steady posture. II. Controllable parameters of the muscle. *Biophysics 11*: 565–578.

Feldman, A.G., and Latash, M.L. (2005). Testing hypotheses and the advancement of science: recent attempts to falsify the equilibrium point hypothesis. *Exp. Brain Res. 161*: 91–103.

Felipe, N.J., and Sommer, R. (1966). Invasions of personal space. *Social Problems 14*: 206–214.

Ferrier, D. (1873). Experimental researches in cerebral physiology and pathology. *West Riding Lunatic Asylum Medical Reports 3*: 30–96.

Ferrier, D. (1874). Experiments on the brain of monkeys – No. 1. *Proc. R. Soc. Lond. 23*: 409–430.

Finkenstadt, T., and Ewert, J.P. (1983). Visual pattern discrimination through interactions of neural networks: a combined electrical brain stimulation, brain lesion, and extracellular recording study in Salamandra salamandra, *J. Comp. Physiol. 153*: 99–110.

Flash, T., and Hogan, N. (1985). The coordination of arm movements: An experimentally confirmed experimental model. *J. Neurosci. 5*: 1688–1703.

Flourens, P. (1960). Investigations of the properties and the functions of the various parts which compose the cerebral mass [Recherches sur la structure de la couche corticale des circonvolutions du cerveau] (Tr. G. von Bonin). In: *Some papers on the cerebral cortex*. Von Bonin, G. (Ed.). Springfield, IL: Charles C Thomas Publisher, pp. 3–21. (Original work published 1824)

Foerster, O. (1936). The motor cortex of man in the light of Hughlings Jackson's doctrines. *Brain 59*: 135–159.

Fogassi, L., Gallese, V., Buccino, G., Craighero, L., Fadiga, L., and Rizzolatti, G. (2001). Cortical mechanism for the visual guidance of hand grasping movements in the monkey: A reversible inactivation study. *Brain 124*: 571–586.

Fogassi, L., Gallese, V., Fadiga, L., Luppino, G., Matelli, M., and Rizzolatti, G. (1996). Coding of peripersonal space in inferior premotor cortex (area F4). *J. Neurophysiol. 76*: 141–157.

Freedman, E.G., Stanford, T.R,, and Sparks, D.L. (1996). Combined eye-head gaze shifts produced by electrical stimulation of the superior colliculus in rhesus monkeys. *J. Neurophysiol. 76*: 927–952.

Fritsch, G., and Hitzig, E. (1960). Uber die elektrishe Erregbarkeit des Grosshirns [On the electrical excitability of the cerebrum]. Tr. G. von Bonin. In: *Some papers on the cerebral cortex*. Von Bonin, G. (Ed.). Springfield, IL: Charles C Thomas Publisher, pp. 73–96. (Original work published 1870 in *Arch. f. Anat., Physiol und wissenchaftl. Mediz., Leipzig*, 300–332)

Fu, Q.G., Suarez, J.I., and Ebner, T.J. (1993). Neuronal specification of direction and distance during reaching movements in the superior precentral premotor area and primary motor cortex of monkeys. *J. Neurophysiol. 70*: 2097–2116.

Fujii, N., Mushiake, H., and Tanji, J. (2000). Rostrocaudal distinction of the dorsal premotor area based on oculomotor involvement. *J. Neurophysiol. 83*: 1764–1769.

Fulton, J. (1934). Forced grasping and groping in relation to the syndrome of the premotor area. *Arch. Neurol. Psychiat. 31*: 221–235.

Fulton, J. (1935). A note on the definition of the "motor" and "premotor" areas. *Brain 58*: 311–316.

Gaffney, G.R., Tsai, L.Y., Kuperman, S., and Minchin, S. (1987). Cerebellar structure in autism. *Am. J. Dis. Child. 141*: 1330–1332.

Galea, M.P., and Darian-Smith, I. (1994). Multiple corticospinal neuron populations in the macaque monkey are specified by their unique cortical origins, spinal terminations, and connections. *Cereb. Cortex 4*: 166–194.

Gallese, V., Fadiga, L., Fogassi, L., and Rizzolatti, G. (1996). Action recognition in the premotor cortex. *Brain 119*: 593–609.

Gallese, V., Keysers, C., and Rizzolatti, G. (2004). A unifying view of the basis of social cognition. *Trends Cogn. Sci. 8*: 396–403.

Galvani, L. (1791). De viribus electricitatis in motu musculari commentarius [Commentary on the effect of electricity on muscular motion]. *De Bononiensi Scientiarum et Artium Instituto atque Academia commentarii 7*: 363–418.

Gaymard, B., Pierrot-Deseilligny, C., and Rivaud, S. (1990). Impairment of sequences of memory-guided saccades after supplementary motor area lesions. *Ann. Neurol. 28*: 622–626.

Gentilucci, M., Fogassi, L., Luppino, G., Matelli, M., Camarda, R., and Rizzolatti, G. (1988). Functional organization of inferior area 6 in the macaque monkey. I. Somatotopy and the control of proximal movements. *Exp. Brain Res. 71*: 475–490.

Gentilucci, M., Scandolara, C., Pigarev, I.N., and Rizzolatti, G. (1983). Visual responses in the postarcuate cortex (area 6) of the monkey that are independent of eye position. *Exp. Brain Res. 50*: 464–468.

Georgopoulos, A.P., Ashe, J., Smyrnis, N., and Taira, M. (1992). The motor cortex and the coding of force. *Science 256*: 1692–1695.

Georgopoulos, A.P., Kalaska, J.F., Caminiti, R., and Massey, J.T. (1982). On the relations between the direction of two-dimensional arm movements and cell discharge in primate motor cortex. *J. Neurosci. 2*: 1527–1537.

Georgopoulos, A.P., Kettner, R.E., and Schwartz, A.B. (1988). Primate motor cortex and free arm movements to visual targets in three-dimensional space. II. Coding of the direction of movement by a neuronal population. *J. Neurosci. 8*: 2928–2937.

Georgopoulos, A.P., Schwartz, A.B., and Kettner, R.E. (1986). Neuronal population coding of movement direction. *Science 233*: 1416–1419.

Gerloff, C., Corwell, B., Chen, R., Hallett, M., and Cohen, L.G. (1997). Stimulation over the human supplementary motor area interferes with the organization of future elements in complex motor sequences. *Brain 120*: 1587–1602.

Gernsbacher, M.A. (2004). Language is more than speech: A case study. *J. Developmental and Learning Disorders 8*: 81–98.

Gierer, A., and Muller, C.M. (1995). Development of layers, maps and modules. *Curr. Opin. Neurobiol. 5*: 91–97.

Giszter, S.F., Mussa-Ivaldi, F.A., and Bizzi, E. (1993). Convergent force fields organized in the frog's spinal cord. *J. Neurosci. 13*: 467–491.

Goldberg, M.C., Lasker, A.G., Zee, D.S., Garth, E., Tien, A., and Landa, R.J. (2002). Deficits in the initiation of eye movements in the absence of a visual target in adolescents with high functioning autism. *Neuropsychologia 40*: 2039–2049.

Gottlieb, J.P., Bruce, C.J., and MacAvoy, M.G. (1993). Smooth eye movements elicited by microstimulation in the primate frontal eye field. *J. Neurophysiol. 69*: 786–799.

Gould, H.J. 3rd, Cusick, C.G., Pons, T.P., and Kaas, J.H. (1986). The relationship of corpus callosum connections to electrical stimulation maps of motor, supplementary motor, and the frontal eye fields in owl monkeys. *J. Comp. Neurol. 247*: 297–325.

Grafton, S.T., Arbib, M.A., Fadiga, L., and Rizzolatti, G. (1996). Localization of grasp representations in humans by positron emission tomography. 2. Observation compared with imagination. *Exp. Brain Res. 112*: 103–111.

Graziano, M.S.A. (2006). The organization of behavioral repertoire in motor cortex. *Ann. Rev. Neurosci. 29*: 105–134.

Graziano, M.S.A., and Aflalo, T.N. (2007), Mapping behavior repertoire onto the cortex. *Neuron 56*: 239–251.

Graziano, M.S.A., Aflalo, T., and Cooke, D.F. (2005). Arm movements evoked by electrical stimulation in the motor cortex of monkeys. *J. Neurophysiol. 94*: 4209–4223.

Graziano, M.S.A., Alisharan, S.A., Hu, X., and Gross, C.G. (2002a). The clothing effect: Tactile neurons in the precental gyrus do not respond to the touch of the familiar primate chair. *Proc. Natl. Acad. Sci. USA 99*: 11930–11933.

Graziano, M.S.A., and Botvinick, M.M. (2002). How the brain represents the body: insights from neurophysiology and psychology. In: *Common mechanisms in perception and action: attention and performance XIX*. Prinz, W., and Hommel, B. (Eds.). Oxford, UK: Oxford University Press, pp. 136–157.

Graziano, M.S.A., and Cooke, D.F. (2006). Parieto-frontal interactions, personal space, and defensive behavior. *Neuropsychologia 44*: 845–859.

Graziano, M.S.A., Cooke, D.F., Taylor, C.S.R., and Moore, T. (2004). Distribution of hand location in monkeys during spontaneous behavior. *Exp. Brain Res. 155*: 30–36.

Graziano, M.S.A., and Gandhi, S. (2000). Location of the polysensory zone in the precentral gyrus of anesthetized monkeys. *Exp. Brain Res. 135*: 259–266.

Graziano, M.S.A., and Gross, C.G. (1998). Spatial maps for the control of movement. *Curr. Opin. Neurobiol. 8*: 195–201.

Graziano, M.S.A., Hu, X.T., and Gross, C.G. (1997a). Visuo-spatial properties of ventral premotor cortex. *J. Neurophysiol. 77*: 2268–2292.

Graziano, M.S.A., Hu, X.T., and Gross, C.G. (1997b). Coding the locations of objects in the dark. *Science 277*: 239–241.

Graziano, M.S.A., Patel, K.T., and Taylor, C.S.R. (2004). Mapping from motor cortex to biceps and triceps altered by elbow angle. *J. Neurophysiol. 92*: 395–407.

Graziano, M.S.A., Reiss, L.A., and Gross, C.G. (1999). A neuronal representation of the location of nearby sounds. *Nature 397*: 428–430.

Graziano, M.S.A., Taylor, C.S.R., and Moore, T. (2002). Complex movements evoked by microstimulation of precentral cortex. *Neuron 34*: 841–851.

Graziano, M.S.A., Yap, G.S., and Gross, C.G. (1994). Coding of visual space by premotor neurons. *Science 266*: 1054–1057.

Gross, C.G. (1997). Emanual Swedenborg: A neuroscientist before his time. *Neuroscientist 3*: 142–147.

Grunbaum, A., and Sherrington, C. (1901). Observations on the physiology of the cerebral cortex of some of the higher apes (Preliminary communication). *Proc. R. Soc. Lond. 69*: 206–209.

Grunbaum, A., and Sherrington, C. (1903). Observations on the physiology of the cerebral cortex of the anthropoid apes. *Proc. R. Soc. Lond. 72*: 152–155.

Guitton, D., Crommelinck, M., and Roucoux, A. (1980). Stimulation of the superior colliculus in the alert cat. I. Eye movements and neck EMG activity evoked when the head is restrained. *Exp. Brain Res. 39*: 63–73.

Haiss, F., and Schwarz, C. (2005). Spatial segregation of different modes of movement control in the whisker representation of rat primary motor cortex. *J. Neurosci. 25*: 1579–1587.

Hall, E.T. (1966). *The hidden dimension*. Garden City, NY: Anchor Books.

Hallett, M., Lebiedowska, M.K., Thomas, S.L., Stanhope, S.J., Denckla, M.B., and Rumsey, J. (1993). Locomotion of autistic adults. *Arch. Neurol. 50*: 1304–1308.

Halsband, U., Matsuzaka, Y., and Tanji, J. (1994). Neuronal activity in the primate supplementary, pre-supplementary and premotor cortex during externally and internally instructed sequential movements. *Neurosci. Res. 20*: 149–155.

Hardan, A.Y., Kilpatrick, M., Keshavan, M.S., and Minshew, N.J. (2003). Motor performance and anatomic magnetic resonance imaging (MRI) of the basal ganglia in autism. *J. Child Neurol. 18*: 317–324.

Hashimoto, T., Tayama, M., Miyazaki, M., Murakawa, K., and Kuroda, Y. (1993). Brainstem and cerebellar vermis involvement in autistic children. *J. Child Neurol. 8*: 149–153.

He, S.Q., Dum, R.P., and Strick, P.L. (1993). Topographic organization of corticospinal projections from the frontal lobe: motor areas on the lateral surface of the hemisphere. *J. Neurosci. 13*: 952–980.

He, S.Q., Dum, R.P., and Strick, P.L. (1995). Topographic organization of corticospinal projections from the frontal lobe: motor areas on the medial surface of the hemisphere. *J. Neurosci. 15*: 3284–3306.

Hediger, H. (1955). *Studies of the psychology and behavior of captive animals in zoos and circuses*. New York: Criterion Books.

Heffner, R., and Masterton, B. (1975). Variation in form of the pyramidal tract and its relationship to digital dexterity. *Brain Behav. Evol. 12*: 161–200.

Heffner, R., and Masterton, B. (1983). The role of the corticospinal tract in the evolution of human digital dexterity. *Brain Behav. Evol. 23*: 165–183.

Hess, W.R. (1957). *Functional organization of the diencephalons*. New York Grune and Stratton.

Hines, M. (1929). On cerebral localization. *Physiol. Rev. 9*: 462–574.

Hitzig, E. (1900). Houghlings Jackson and the cortical motor centres in the light of physiological research. *Brain 23*: 545–581.

Hocherman, S., and Wise, S.P. (1991). Effects of hand movement path on motor cortical activity in awake, behaving rhesus monkeys. *Exp. Brain Res. 83*: 285–302.

Hoebel, B.G. (1969). Feeding and self-stimulation. *Ann. NY Acad. Sci. 157*: 758–778.

Hoff, E.C., and Hoff, H.E. (1934). Spinal terminations of the projection fibres from the motor cortex of primates. *Brain 57*: 454–474.

Hoffman, J.E., and Subramaniam, B. (1995). The role of visual attention in saccadic eye movements. *Percept. Psychophys. 57*: 787–795.

Holdefer, R.N., and Miller, L.E. (2002). Primary motor cortical neurons encode functional muscle synergies. *Exp. Brain Res. 146*: 233–243.

Horowitz, M.J., Duff, D.F., and Stratton, L.O. (1964). Body-buffer zone: exploration of personal space. *Arch. Gen. Psychiat. 11*: 651–656.

Horsley, V., and Schaffer, E.A. (1888). A record of experiments upon the functions of the cerebral cortex. *Phil. Trans. 179*: 1–45.

Huang, C.S., Hiraba, H., Murray, G.M., and Sessle, B.J. (1989). Topographical distribution and functional properties of cortically induced rhythmical jaw movements in the monkey (Macaca fascicularis). *J. Neurophysiol.* 61: 635–650.

Hubel, D., and Wiesel, T. (1962). Receptive fields, binocular interaction and functional architecture in the cat's visual cortex. *J. Physiol.* 160: 106–154.

Huntley, G.W., and Jones, E.G. (1991). Relationship of intrinsic connections to forelimb movement representations in monkey motor cortex: a correlative anatomic and physiological study. *J. Neurophysiol.* 66: 390–413.

Iacoboni, M., and Dapretto, M. (2006). The mirror neuron system and the consequences of its dysfunction. *Nat. Rev. Neurosci.* 7: 942–951.

Iacoboni, M., Woods, R.P., Brass, M., Bekkering, H., Mazziotta, J.C., and Rizzolatti, G. (1999). Cortical mechanisms of human imitation. *Science 286*: 2526–2528.

Ingudomnukul, E., Baron-Cohen, S., Wheelwright, S., and Knickmeyer, R. (2007). Elevated rates of testosterone-related disorders in women with autism spectrum conditions. *Horm. Behav.* 51: 597–604.

Jackson, J.H. (1870). Study of convulsions (St. Andrews Reports Vol. 3). In: *Selected writings of John Houghlings Jackson*, Vol. 1. Taylor, J. (Ed.). London: Hodder and Stoughton, pp. 8–36.

Jackson, J.H. (1875). On the anatomical and physiological localization of movements in the brain. In: *Selected writings of John Houghlings Jackson*, Vol. 1. Taylor, J. (Ed.). London: Hodder and Stoughton, pp. 37–76.

Jackson, J.H. (1890). On convulsive seizures. *Lancet 1*: 685–688, 735–738, 785–788.

Jankowska, E., Padel, Y., and Tanaka, R. (1975). The mode of activation of pyramidal tract cells by intracortical stimuli. *J. Physiol.* 249: 617–636.

Jeannerod, M. (1986). The formation of finger grip during prehension. A cortically mediated visuomotor pattern. *Behav. Brain Res.* 19: 99–116.

Jenkins, I.H., Brooks, D.J., Nixon, P.D., Frackowiak, R.S., and Passingham, R.E. (1994). Motor sequence learning: a study with positron emission tomography. *J. Neurosci.* 14: 3775–3790.

Johnson, P.B., Ferraina, S., Bianchi, L., and Caminiti, R. (1996). Cortical networks for visual reaching: physiological and anatomical organization of frontal and parietal lobe arm regions. *Cereb. Cortex 6*: 102–119.

Kaas, J.H., and Catania, K.C. (2002). How do features of sensory representations develop? *Bioessays 24*: 334–343.

Kakei, S., Hoffman, D., and Strick, P. (1999). Muscle and movemet representations in the primary motor cortex. *Science 285*: 2136–2139.

Kakei, S., Hoffman, D., and Strick, P. (2001). Direction of action is represented in the ventral premotor cortex. *Nat. Neurosci.* 4: 969–970.

Kanner, L. (1943). Autistic disturbances of affective contact. *Nervous Child 2*: 217–250.

Karni, A., Meyer, G., Jezzard, P., Adams, M.M., Turner, R., and Ungerleider, L.G. (1995). Functional MRI evidence for adult motor cortex plasticity during motor skill learning. *Nature 377*: 155–158.

Kastner, S., DeSimone, K., Konen, C.S., Szczepanski, S.M., Weiner, K.S., and Schneider, K.A. (2007). Topographic maps in human frontal cortex revealed in memory-guided saccade and spatial working-memory tasks. *J. Neurophysiol.* 97: 3494–3507.

Kaufmann, W.E., Cooper, K.L., Mostofsky, S.H., Capone, G.T., Kates, W.R., Newschaffer, C.J., Bukelis, I., Stump, M.H., Jann, A.E., and Lanham, D.C. (2003). Specificity of cerebellar vermian abnormalities in autism: a quantitative magnetic resonance imaging study. *J. Child Neurol.* 18: 463–470.

Kellaway, P. (1946). The part played by electric fish in the early history of bioelectricity and electrotherapy. *Bull. Hist. Med. 20*: 112–137.

Kennard, M.A. (1935). Corticospinal fibres arising in the premotor area of the monkey. *Arch. Neurol. Psychiat. 33*: 698–711.

Kettner, R.E., Schwartz, A.B., and Georgopoulos, A.P. (1988). Primate motor cortex and free arm movements to visual targets in three-dimensional space. III. Positional gradients and population coding of movement direction from various movement origins. *J. Neurosci. 8*: 2938–2947.

Kimmel, D.L., and Moore, T. (2007). Temporal patterning of saccadic eye movement signals. *J. Neurosci. 27*: 7619–7630.

King, M.B., and Hoebel, B.G. (1968). Killing elicited by brain stimulation in rats. *Comm. Behav. Biol. 2*: 173–177.

Kleim, J.A., Barbay, S., and Nudo, R.J. (1998). Functional reorganization of the rat motor cortex following motor skill learning. *J. Neurophysiol. 80*: 3321–3325.

Kleinschmidt, A., Nitschke, M.F., and Frahm, J. (1997). Somatotopy in the human motor cortex hand area. A high-resolution functional MRI study. *Eur. J. Neurosci. 9*: 2178–2186.

Knight, T.A., and Fuchs, A.F. (2007). Contribution of the frontal eye field to gaze shifts in the head-unrestrained monkey: effects of microstimulation. *J. Neurophysiol. 97*: 618–634.

Kohen-Raz, R., Volkmar, F.R., and Cohen, D.J. (1992). Postural control in children with autism. *J. Autism Dev. Disord. 22*: 419–432.

Kohonen, T. (1982). Self-organizing formation of topologically correct feature maps. *Biol. Cybern. 43*: 59–69.

Kohonen, T. (2001). *Self-organizing maps*. Berlin, Germany: Springer.

Kowler, E., Anderson, E., Dosher, B., and Blaser, E. (1995). The role of attention in the programming of saccades. Vis. Res. 35: 1897–1916.

Kurylo, D.D., and Skavenski, A.A. (1991). Eye movements elicited by electrical stimulation of area PG in the monkey. *J. Neurophysiol. 65*: 1243–1253.

Kuypers, H.G.J.M., and Brinkman, J. (1970). Precental projections to different parts of the spinal intermediate zone in the rhesus monkey. *Brain Res. 24*: 29–48.

Kwan, H.C., MacKay, W.A., Murphy, J.T., and Wong, Y.C. (1978). Spatial organization of precentral cortex in awake primates. II. Motor outputs. *J. Neurophysiol. 41*: 1120–1131.

Lan, N., and Crago, P.E. (1994). Optimal control of antagonistic muscle stiffness during voluntary movements. *Biol. Cybern. 71*: 123–135.

Landgren, S., Phillips, C.G., and Porter, R. (1962). Cortical fields of origin of the monosynaptic pyramidal pathways to some alpha motoneurones of the baboon's hand and forearm. *J. Physiol. 161*: 112–125.

Landis, C., and Hunt, W.A. (1939). *The startle pattern*. New York: Farrar and Rinehart Inc.

Lassek, A.M. (1941). The pyramidal tract of the monkey. *J. Comp. Neurol. 74*: 193–202.

Leary, M.R., and Hill, D.A. (1996). Moving on: autism and movement disturbance. *Ment. Retard. 34*: 39–53.

Lee, D., and Quessy, S. (2003). Activity in the supplementary motor area related to learning and performance during a sequential visuomotor task. *J. Neurophysiol. 89*: 1039–1056.

Lemon, R.N., Johansson, R.S., and Westling, G. (1995). Corticospinal control during reach, grasp, and precision lift in man. *J. Neurosci. 15*: 6145–6156.

Lewis, J.W., and Van Essen, D.C. (2000). Corticocortical connections of visual, sensorimotor, and multimodal processing areas in the parietal lobe of the macaque monkey. *J. Comp. Neurol. 428*: 112–137.

Li, W., Todorov, E., and Pan, X. (2005). Hierarchical feedback and learning for multijoint arm movement control. *Conf. Proc. IEEE Eng. Med. Biol. Soc. 4*: 4400–4403.

Liberman, A.M., Cooper, F.S., Shankweiler, D.P., and Studdert-Kennedy, M. (1967). Perception of the speech code. *Psychol. Rev. 74*: 431–461.

Liu, D., and Todorov, E. (2007). Evidence for the flexible sensorimotor strategies predicted by optimal feedback control. *J. Neurosci. 27*: 9354–9368.

Loeb, E.P., Giszter, S.F., Borghesani, P., and Bizzi, E. (1993). Effects of dorsal root cut on the forces evoked by spinal microstimulation in the spinalized frog. *Somatosens. Mot. Res. 10*: 81–95.

Logothetis, N., Sultan, F., Murayama, Y., Augath, M., Steudel, T., and Oeltermann, A. (2006). Microstimulation and fMRI in anesthetized and alert monkeys: conditions for transsynaptic BOLD activation. *Soc. Neurosci. Abstr.* 114.10.

Lu, X., and Ashe, J. (2005). Anticipatory activity in primary motor cortex codes memorized movement sequences. *Neuron 45*: 967–973.

Lu, M.T., Preston, J.B., and Strick, P.L. (1994). Interconnections between the prefrontal cortex and the premotor areas in the frontal lobe. *J. Comp. Neurol. 341*: 375–392.

Luppino, G., Matelli, M., Camarda, R.M., Gallese, V., and Rizzolatti, G. (1991). Multiple representations of body movements in mesial area 6 and the adjacent cingulate cortex: an intracortical microstimulation study in the macaque monkey. *J. Comp. Neurol. 311*: 463–482.

Luppino, G., Matelli, M., Camarda, R., and Rizzolatti, G. (1993). Corticocortical connections of area F3 (SMA-proper) and area F6 (pre-SMA) in the macaque monkey. *J. Comp. Neurol. 338*: 114–140.

Luppino, G., Murata, A., Govoni, P., and Matelli, M. (1999). Largely segregated parietofrontal connections linking rostral intraparietal cortex (areas AIP and VIP) and the ventral premotor cortex (areas F5 and F4). *Exp. Brain Res. 128*: 181–187.

Macpherson, J., Marangoz, C., Miles, T.S., and Wiesendanger, M. (1982). Microstimulation of the supplementary motor area (SMA) in the awake monkey. *Exp. Brain Res. 45*: 410–416.

Macpherson, J., Wiesendanger, M., Marangoz, C., and Miles, T.S. (1982). Corticospinal neurones of the supplementary motor area of monkeys. A single unit study. *Exp. Brain Res. 48*: 81–88.

Macrae, C.N., Hood, B.M., Milne, A.B., Rowe, A.C., and Mason, M.F. (2002). Are you looking at me? Eye gaze and person perception. *Psychol. Sci. 13*: 460–464.

Maier, M.A., Olivier, E., Baker, S.N., Kirkwood, P.A., Morris, T., and Lemon, R.N. (1997). Direct and indirect corticospinal control of arm and hand motoneurons in the squirrel monkey (Saimiri sciureus). *J. Neurophysiol. 78*: 721–733.

Maier, M.A., Shupe, L.E., and Fetz, E.E. (2005). Dynamic neural network models of the premotoneuronal circuitry controlling wrist movements in primates. *J. Comput. Neurosci. 19*: 125–146.

Mari, M., Castiello, U., Marks, D., Marraffa, C., and Prior, M. (2003). The reach-to-grasp movement in children with autism spectrum disorder. *Phil. Trans. R. Soc. Lond. B Biol. Sci. 358*: 393–403.

Martin, J.H., Engber, D., and Meng, Z. (2005). Effect of forelimb use on postnatal development of the forelimb motor representation in primary motor cortex of the cat. *J. Neurophysiol. 93*: 2822–2831.

Martinez-Trujillo, J.C., Wang, H., and Crawford, J.D. (2003). Electrical stimulation of the supplementary eye fields in the head-free macaque evokes kinematically normal gaze shifts. *J. Neurophysiol. 89*: 2961–2974.

Marzke, M.W., and Marzke, R.F. (2000). Evolution of the human hand: approaches to acquiring, analysing and interpreting the anatomical evidence. *J. Anat. 197*: 121–140.

Matelli, M., and Luppino, G. (2001). Parietofrontal circuits for action and space perception in the macaque monkey. *Neuroimage 14*: S27–S32.

Matelli, M., Luppino, G., and Rizzolatti, G. (1985). Patterns of cytochrome oxidase activity in the frontal agranular cortex of the macaque monkey. *Behav. Brain Res. 18*: 125–136.

Matelli, M., Luppino, G., and Rizzolatti, G. (1991). Architecture of superior and mesial area 6 and the adjacent cingulate cortex in the macaque monkey. *J. Comp. Neurol. 311*: 445–462.

Matsuzaka, Y., Aizawa, H., and Tanjii, J. (1992). A motor area rostral to the supplementary motor area (presupplementary motor area) in the monkey: neuronal activity during a learned motor task. *J. Neurophysiol. 68*: 653–662.

Meier, J.D., Aflalo, T.N.S., Kastner, S., and Graziano, M.S.A. (2007). Complex somatotopic organization in human motor cortex. *Soc. Neurosci. Abs.* 292.22.

Mennie, N., Hayhoe, M., and Sullivan, B. (2007). Look-ahead fixations: anticipatory eye movements in natural tasks. *Exp. Brain Res. 179*: 427–442.

Messier, J., and Kalaska, J.F. (2000). Covariation of primate dorsal premotor cell activity with direction and amplitude during a memorized-delay reaching task. *J. Neurophysiol. 84*:152–165.

Minshew, N.J., Luna, B., and Sweeney, J.A. (1999). Oculomotor evidence for neocortical systems but not cerebellar dysfunction in autism. *Neurology 52*: 917–922.

Minshew, N.J., Sung, K., Jones, B.L., and Furman, J.M. (2004). Underdevelopment of the postural control system in autism. *Neurology 63*: 2056–2061.

Mitz, A.R., and Wise, S.P. (1987). The somatotopic organization of the supplementary motor area: intracortical microstimulation mapping. *J. Neurosci. 7*: 1010–1021.

Moore, T., and Armstrong, K.M. (2003). Selective gating of visual signals by microstimulation of frontal cortex. *Nature 421*: 370–373.

Moore, T., and Fallah, M. (2004). Microstimulation of the frontal eye field and its effects on covert spatial attention. *J. Neurophysiol. 91*: 152–162.

Moran, D.W., and Schwartz, A.B., (1999). Motor cortical representation of speed and direction during reaching. *J. Neurophysiol. 82*: 2676–2692.

Muakkassa, K.F., and Strick, P.L. (1979). Frontal lobe inputs to primate motor cortex: evidence for four somatotopically organized "premotor" areas. *Brain Res. 177*: 176–182.

Muhammad, R., Wallis, J.D., and Miller, E.K. (2006). A comparison of abstract rules in the prefrontal cortex, premotor cortex, inferior temporal cortex, and striatum. *J. Cog. Neurosci. 18*: 974–989.

Muller, J.R., Philiastides, M.G., and Newsome, W.T. (2005). Microstimulation of the superior colliculus focuses attention without moving the eyes. *Proc. Natl. Acad. Sci. USA 102*: 524–529.

Murata, A., Fadiga, L., Fogassi, L., Gallese, V., Raos, V., and Rizzolatti, G. (1997). Object representation in the ventral premotor cortex (area F5) of the monkey. *J. Neurophysiol. 78*: 2226–2230.

Murray, E.A., and Coulter, J.D. (1981). Organization of corticospinal neurons in the monkey. *J. Comp. Neurol. 195*: 339–365.

Mushiake, H., Inase, M., and Tanjii, J. (1990). Selective coding of motor sequence in the supplementary motor area of the monkey cerebral cortex. *Exp. Brain Res. 82*: 208–210.

Napier, J.R. (1956). The prehensile movements of the human hand. *J. Bone Joint Surg. 38B*: 902–913.

Nudo, R.J., and Masterton, R.B. (1990). Descending pathways of the spinal cord, III: Sites of origin of the corticospinal tract. *J. Comp. Neurol. 296*: 559–583.

Nudo, R.J., and Milliken, G.W. (1996). Reorganization of movement representations in primary motor cortex following focal ischemic infarcts in adult squirrel monkeys. *J. Neurophys. 75*: 2144–2149.

Nudo, R.J., Milliken, G.W., Jenkins, W.M., and Merzenich, M.M. (1996). Use-dependent alterations of movement representations in primary motor cortex of adult squirrel monkeys. *J. Neurosci. 16*: 785–807.

O'Leary, D.D., and McLaughlin, T. (2005). Mechanisms of retinotopic map development: Ephs, ephrins, and spontaneous correlated retinal activity. *Prog. Brain Res. 147*: 43–65.

Olds, J., and Milner, P. (1954). Positive reinforcement produced by electrical stimulation of septal area and other regions of rat brain. *J. Comp. Physiol. Psychol. 47*: 419–427.

Paninski, L., Fellows, M.R., Hatsopoulos, N.G., and Donoghue, J.P. (2004). Spatiotemporal tuning of motor cortical neurons for hand position and velocity. *J. Neurophysiol. 91*: 515–532.

Park, M.C., Belhaj-Saif, A., and Cheney, P.D. (2004). Properties of primary motor cortex output to forelimb muscles in rhesus macaques. *J. Neurophysiol. 92*: 2968–2984.

Park, M.C., Belhaj-Saif, A., Gordon, M., and Cheney, P.D. (2001). Consistent features in the forelimb representation of primary motor cortex in rhesus macaques. *J. Neurosci. 21*: 2784–2792.

Pascual-Leone, A., Nguyet, D., Cohen, L.G., Brasil-Neto, J.P., Cammarota, A., and Hallett, M. (1995). Modulation of muscle responses evoked by transcranial magnetic stimulation during the acquisition of new fine motor skills. *J. Neurophys. 74*: 1037–1045.

Passingham, R.E. (1985). Premotor cortex: sensory cues and movement. *Behav. Brain Res. 18*: 175–185.

Passingham, R.E. (1986). Cues for movement in monkeys (Macaca mulatta) with lesions in premotor cortex. *Behav. Neurosci. 100*: 695–703.

Penfield, W. (1959). The interpretive cortex. *Science 129*: 1719–1725.

Penfield, W., and Boldrey, E. (1937). Somatic motor and sensory representation in the cerebral cortex of man as studied by electrical stimulation. *Brain 60*: 389–443.

Penfield, W., and Rasmussen, T. (1950). *The cerebral cortex of man. a clinical study of localization of function.* New York: Macmillan.

Penfield, W., and Welch, K. (1951). The supplementary motor area of the cerebral cortex: A clinical and experimental study. *Am. Med. Ass. Arch. Neurol. Psychiat. 66*: 289–317.

Picard, N., and Strick, P.L. (2003). Activation of the supplementary motor area (SMA) during performance of visually guided movements. *Cereb. Cortex 13*: 977–986.

Piven, J., Nehme, E., Simon, J., Barta, P., Pearlson, G., and Folstein, S.E. (1992). Magnetic resonance imaging in autism: measurement of the cerebellum, pons, and fourth ventricle. *Biol. Psychiat. 31*: 491–504.

Polit, A., and Bizzi, E. (1979). Characteristics of motor programs underlying arm movements in monkeys. *J. Neurophysiol. 42*: 183–194.

Preuschoft, S. (1992). "Laughter" and "smile" in Barbary macaques (Macaca sylvanus). *Ethology 91*: 220–236.

Preuss, T.M., Stepniewska, I., and Kaas, J.H. (1996). Movement representation in the dorsal and ventral premotor areas of owl monkeys: a microstimulation study. *J. Comp. Neurol. 371*: 649–676.

Ramanathan, D., Conner, J.M., and Tuszynski, M.H. (2006). A form of motor cortical plasticity that correlates with recovery of function after brain injury. *Proc. Natl. Acad. Sci. USA 103*: 11370–11375.

Ranck, J.B. (1974). Which elements are excited in electrical stimulation of mammalian central nervous system: a review. *Brain Res. 98*: 417–440.

Raos, V., Umilta, M.A., Murata, A., Fogassi, L., and Gallese, V. (2006). Functional properties of grasping-related neurons in the ventral premotor area F5 of the macaque monkey. *J. Neurophysiol. 95*: 709–729.

Rathelot, J.A., and Strick, P.L. (2006). Muscle representation in the macaque motor cortex: an anatomical perspective. *Proc. Natl. Acad. Sci. USA 103*: 8257–8262.

Reina, G.A., Moran, D.W., and Schwartz, A.B. (2001). On the relationship between joint angular velocity and motor cortical discharge during reaching. *J. Neurophysiol. 85*: 2576–2589.

Reynolds, E.H. (2004). Todd, Faraday, and the electric basis of epilepsy. *Epilepsia 45*: 985–992.

Richter, C., and Hines, M. (1932). The production of the "grasp reflex" in adult macaques by experimental frontal lobe lesions. *Proc. Assoc. Res. Nerv. Ment. Dis. 13*: 211–224.

Ritvo, E.R., Freeman, B.J., Scheibel, A.B., Duong, T., Robinson, H., Guthrie, D., and Ritvo, A. (1986). Lower Purkinje cell counts in the cerebella of four autistic subjects: initial findings of the UCLA-NSAC Autopsy Research Report. *Am. J. Psychiat. 143*: 862–866.

Rizzolatti, G., and Arbib, M.A. (1998). Language within our grasp. *Trends Neurosci. 21*: 188–194.

Rizzolatti, G., Camarda, R., Fogassi, L., Gentilucci, M., Luppino, G., and Matelli, M. (1988). Functional organization of inferior area 6 in the macaque monkey. II. Area F5 and the control of distal movements. *Exp. Brain Res. 71*: 491–507.

Rizzolatti, G., and Craighero, L. (2004). The mirror-neuron system. *Ann. Rev. Neurosci. 27*: 169–192.

Rizzolatti, G., and Luppino, G. (2001). The cortical motor system. *Neuron 31*: 889–901.

Rizzolatti, G., Riggio, L., Dascola, I., and Umilta, C. (1987). Reorienting attention across the horizontal and vertical meridians: evidence in favor of a premotor theory of attention. *Neuropsychologia 25*: 31–40.

Rizzolatti, G., Scandolara, C., Matelli, M., and Gentilucci, M. (1981). Afferent properties of periarcuate neurons in macaque monkeys. II. Visual responses. *Behav. Brain Res. 2*: 147–163.

Robinson, D.A. (1972). Eye movements evoked by collicular stimulation in the alert monkey. *Vis. Res. 12*: 1795–1808.

Robinson, D.A., and Fuchs, A.F. (1969). Eye movements evoked by stimulation of the frontal eye fields. *J. Neurophysiol. 32*: 637–648.

Rogers, S.J., Hepburn, S.L., Stackhouse, T., and Wehner, E. (2003). Imitation performance in toddlers with autism and those with other developmental disorders. *J. Child Psychol. Psychiat. 44*: 763–781.

Roland, P.E., and Larsen, B. (1976). Focal increase of cerebral blood flow during stereognostic testing in man. *Arch. Neurol. 33*: 551–558.

Roland, P.E., Larsen, B., Lassen, N.A., and Skinhoj, E. (1980). Supplementary motor area and other cortical areas in organization of voluntary movements in man. *J. Neurophysiol. 43*: 118–136.

Roland, P.E., Skinhoj, E., Lassen, N.A., and Larsen, B. (1980). Different cortical areas in man in organization of voluntary movements in extrapersonal space. *J. Neurophysiol. 43*: 137–150.

Romo, R., Hernandez, A., Zainos, A., and Salinas, E. (1998). Somatosensory discrimination based on cortical microstimulation. *Nature 392*: 387–390.

Rosa, M.G., and Tweedale, R. (2005). Brain maps, great and small: lessons from comparative studies of primate visual cortical organization. *Phil. Trans. R. Soc. Lond. B Biol. Sci. 360*: 665–691.

Rosenbaum, D.A., Loukopoulos, L.D., Meulenbroek, R.G., Vaughan, J., and Engelbrecht, S.E. (1995). Planning reaches by evaluating stored postures. *Psychol. Rev. 102*: 28–67.

Russo, G.S., and Bruce, C.J. (2000). Supplementary eye field: representation of saccades and relationship between neural response fields and elicited eye movements. *J. Neurophysiol. 84*: 2605–2621.

Saarinen, J., and Kohonen, T. (1985).. Self-organized formation of colour maps in a model cortex. *Perception 14*: 711–719.

Salzman, C.D., Britten, K.H., and Newsome, W.T. (1990). Cortical microstimulation influences perceptual judgements of motion direction. *Nature 346*: 174–177.

Sanes, J.N., Donoghue, J.P., Thangaraj, V., Edelman, R.R., and Warach, S. (1995). Shared neural substrates controlling hand movements in human motor cortex. *Science 268*: 1775–1777.

Sanes, J.N., Wang, J., and Donoghue, J.P. (1992). Immediate and delayed changes of rat cortical output representation with new forelimb configurations. *Cereb. Cortex 2*: 141–152.

Schaafsma, S.J., and Duysens, J. (1996). Neurons in the ventral intraparietal area of awake macaque monkey closely resemble neurons in the dorsal part of the medial superior temporal area in their responses to optic flow patterns. *J. Neurophysiol. 76*: 4056–4068.

Schieber, M.H., and Hibbard, L.S. (1993). How somatotopic is the motor cortex hand area? *Science 261*: 489–492.

Schiff, W. (1965). Perception of impending collision: A study of visually directed avoidant behavior. *Psychological Monographs: General and Applied 79*: 1–26.

Schiff, W., Caviness, J.A., and Gibson, J.J. (1962). Persistent fear responses in rhesus monkeys to the optical stimulus of "looming." *Science 136*: 982–983.

Schiller, P.H., and Stryker, M. (1972). Single-unit recording and stimulation in superior colliculus of the alert rhesus monkey. *J. Neurophysiol. 35*: 915–924.

Schlack, A., Sterbing, S., Hartung, K., Hoffmann, K.P., and Bremmer, F. (2005). Multisensory space representations in the Macaque ventral intraparietal area (VIP). *J. Neurosci. 25*: 4616–4625.

Schlag, J., and Schlag-Rey, M. (1987). Evidence for a supplementary eye field. *J. Neurophysiol. 57*: 179–200.

Schneider, C., Devanne, H., Lavoie, B.A., and Capaday, C. (2002). Neural mechanisms involved in the functional linking of motor cortical points. *Exp. Brain Res. 146*: 86–94.

Schwartz, A.B., Kettner, R.E., and Georgopoulos, A.P. (1988). Primate motor cortex and free arm movements to visual targets in three-dimensional space. I. Relations between single cell discharge and direction of movement. *J. Neurosci.* 8: 2913–2927.

Scott, S.H., and Kalaska, J.F. (1995). Changes in motor cortex activity during reaching movements with similar hand paths but different arm postures. *J. Neurophysiol.* 73: 2563–2567.

Scott, S.H., and Kalaska, J.F. (1997). Reaching movements with similar hand paths but different arm orientations. I. Activity of individual cells in motor cortex. *J. Neurophysiol.* 77: 826–852.

Seidemann, E., Arieli, A., Grinvald, A., and Slovin, H. (2002). Dynamics of depolarization and hyperpolarization in the frontal cortex and saccade goal. *Science* 295: 862–865.

Sergio, L.E., and Kalaska, J.F. (2003). Systematic changes in motor cortex cell activity with arm posture during directional isometric force generation. *J. Neurophysiol.* 89: 212–228.

Serrien, D.J., Strens, L.H., Oliviero, A., and Brown, P. (2002). Repetitive transcranial magnetic stimulation of the supplementary motor area (SMA) degrades bimanual movement control in humans. *Neurosci. Lett.* 328: 89–92.

Sessle, B.J., and Wiesendanger, M. (1982). Structural and functional definition of the motor cortex in the monkey (Macaca fascicularis). *J. Physiol.* 323: 245–265.

Shadmehr, R. (1993). Control of equilibrium position and stiffness through postural modules. *J. Mot. Behav.* 25: 228–241.

Shadmehr, R., and Moussavi, Z.M. (2000). Spatial generalization from learning dynamics of reaching movements. *J. Neurosci.* 20: 7807–7815.

Shepherd, M., Findlay, J.M., and Hockey, R.J. (1986). The relationship between eye movements and spatial attention. *Q. J. Exp. Psychol. A.* 38: 475–491.

Sherrington, C.S. (1939). On the motor area of the cerebral cortex. In: *Selected writings of Sir Charles Sherrington*. Denny-Brown, D. (Ed.). London: Hamish Hamilton Medical Books, pp. 397–439.

Shibutani, H., Sakata, H., and Hyvarinen, J. (1984). Saccade and blinking evoked by microstimulation of the posterior parietal association cortex of the monkey. *Exp. Brain Res.* 55: 1–8.

Shik, M.L., Severin, F.V., and Orlovsky, G.N. (1969). Control of walking and running by means of electrical stimulation of the mesencephalon. *Electroencephalogr. Clin. Neurophysiol.* 26: 549.

Shimazu, H., Maier, M.A., Cerri, G., Kirkwood, P.A., and Lemon, R.N. (2004). Macaque ventral premotor cortex exerts powerful facilitation of motor cortex outputs to upper limb motoneurons. *J. Neurosci.* 24: 1200–1211.

Slovin, H., Strick, P., Hildesheim, R., and Grinvald, A. (2003). Voltage sensitive dye imaging in the motor cortex I. Intra- and intercortical connectivity revealed by microstimulation in the awake monkey. *Soc. Neurosci. Abs.* 554.8.

Smith, B. (1998). *Moving 'em*. Kamuela, HI: Graziers Hui Publisher.

Snyder, L.H., Batista, A.P., and Andersen, R.A. (1997). Coding of intention in the posterior parietal cortex. *Nature* 386: 167–170.

Sommer, R. (1959). Studies in personal space. *Sociometry* 22: 247–260.

Sommer, M.A. and Wurtz, R.H. (2002). A pathway in primate brain for internal monitoring of movements. *Science* 296: 1480–1482.

Stanford, T.R., Freedman, E.G., and Sparks, D.L. (1996). Site and parameters of microstimulation: evidence for independent effects on the properties of saccades evoked from the primate superior colliculus. *J. Neurophysiol.* 76: 3360–3381.

Stepniewska, I., Fang, P.C., and Kaas, J.H. (2005). Microstimulation reveals specialized subregions for different complex movements in posterior parietal cortex of prosimian galagos. *Proc. Natl. Acad. Sci. USA 102*: 4878–4883.

Stone, W.L., Ousley, O.Y., and Littleford, C.D. (1997). Motor imitation in young children with autism: what's the object? *J. Abnorm. Child Psychol. 25*: 475–485.

Strauss, H. (1929). Das Zusammenschrecken [The Startle]. *Journal fur Psychologie und Neurologie 39*: 111–231.

Strick, P.L. (2002). Stimulating research on motor cortex. *Nat. Neurosci. 5*: 714–715.

Takada, M., Nambu, A., Hatanaka, N., Tachibana, Y., Miyachi, S., Taira, M., and Inase, M. (2004). Organization of prefrontal outflow toward frontal motor-related areas in macaque monkeys. *Eur. J. Neurosci. 19*: 3328–3342.

Takarae, Y., Minshew, N.J., Luna, B., Krisky, C.M., and Sweeney, J.A. (2004a). Pursuit eye movement deficits in autism. *Brain 127*: 2584–2594.

Takarae, Y., Minshew, N.J., Luna, B., Krisky, C.M., and Sweeney, J.A. (2004b). Oculomotor abnormalities parallel cerebellar histopathology in autism. *J. Neurol. Neurosurg. Psychiatry 75*: 1359–1361.

Taub, E., Goldberg, I.A., and Taub, P. (1975). Deafferentation in monkeys: pointing at a target without visual feedback. *Exp. Neurol. 46*: 178–186.

Taub, E., Perrella, P.N., and Barro, G. (1973). Behavioral development after forelimb deafferentation on day of birth in monkeys with and without blinding. *Science 181*: 959–960.

Taylor, C.S.R., Cooke, D.F., and Graziano, M.S.A. (2002). Complex mapping from precentral cortex to muscles. *Soc. Neurosci. Abs.* 61.12.

Tehovnik, E.J. (1996). Electrical stimulation of neural tissue to evoke behavioral responses. *J. Neurosci. Methods 65*: 1–17.

Tehovnik, E.J., and Lee, K. (1993). The dorsomedial frontal cortex of the rhesus monkey: topographic representation of saccades evoked by electrical stimulation. *Exp. Brain Res. 96*: 430–442.

Tehovnik, E.J., Slocum, W.M., Carvey, C.E., and Schiller, P.H. (2005). Phosphene induction and the generation of saccadic eye movements by striate cortex. *J. Neurophysiol. 93*: 1–19.

Tehovnik, E.J., Tolias, A.S., Sultan, F., Slocum, W.M., and Logothetis, N.K. (2006). Direct and indirect activation of cortical neurons by electrical microstimulation. *J. Neurophysiol. 96*: 512–521.

Tehovnik, E.J., and Yeomans, J.S. (1987). Circling elicited from the anteromedial cortex and medial pons: refractory periods an summation. *Brain Res. 407*: 240–252.

Teitelbaum, O., Benton, T., Shah, P.K., Prince, A., Kelly, J.L., and Teitelbaum, P. (2004). Eshkol-Wachman movement notation in diagnosis: the early detection of Asperger's syndrome. *Proc. Natl. Acad. Sci. USA 101*: 11909–11914.

Teitelbaum, P., Teitelbaum, O., Nye, J., Fryman, J., and Maurer, R.G. (1998). Movement analysis in infancy may be useful for early diagnosis of autism. *Proc. Natl. Acad. Sci. USA 95*: 13982–13987.

Thier, P., and Andersen, R.A. (1998). Electrical microstimulation distinguishes distinct saccade-related areas in the posterior parietal cortex. *J. Neurophysiol. 80*: 1713–1735.

Ting, L.H., and Macpherson, J.M. (2005). A limited set of muscle synergies for force control during a postural task. *J. Neurophysiol. 93*: 609–613.

Todd, R.B. (1849). On the pathology and treatment of convulsive diseases. Reprinted in 2005, *Epilepsia 46*: 995–1009.

Todorov, E. (2000). Direct cortical control of muscle activation in voluntary arm movements: a model. *Nat. Neurosci. 3*: 391–398.

Todorov, E., and Jordan, M.I. (2002). Optimal feedback control as a theory of motor coordination. *Nat. Neurosci.* 5: 1226–1235.

Tolias, A.S., Sultan, F., Augath, M., Oeltermann, A., Tehovnik, E.J., Schiller, P.H., and Logothetis, N.K. (2005). Mapping cortical activity elicited with electrical microstimulation using fMRI in the macaque. *Neuron* 48: 901–911.

Torres-Oviedo, G., and Ting, L.H. (2007). Muscle synergies characterizing human postural responses. *J. Neurophys.* 98: 2144–2156.

Townsend, J., Harris, N.S., and Courchesne, E. (1996). Visual attention abnormalities in autism: delayed orienting to location. *J. Int. Neuropsychol. Soc.* 2: 541–550.

Townsend, B.R., Paninski, L., and Lemon, R.N. (2006). Linear encoding of muscle activity in primary motor cortex and cerebellum. *J. Neurophysiol.* 96: 2578–2592.

Toyoshima, K., and Sakai, H. (1982). Exact cortical extent of the origin of the corticospinal tract (CST) and the quantitative contribution to the CST in different cytoarchitectonic areas. A study with horseradish peroxidase in the monkey. *J. Hirnforsch.* 23: 257–269.

Tresch, M.C., and Bizzi, E. (1999). Responses to spinal microstimulation in the chronically spinalized rat and their relationship to spinal systems activated by low threshold cutaneous stimulation. *Exp. Brain Res.* 129: 401–416.

Tresch, M.C., Saltiel, P., and Bizzi, E. (1999). The construction of movement by the spinal cord. *Nat. Neurosci.* 2: 162–167.

Vercher, J.L., Sarès, F., Blouin, J., Bourdin, C., and Gauthier, G. (2003). Role of sensory information in updating internal models of the effector during arm tracking. *Prog. Brain Res.* 142: 203–222.

Vilensky, J.A., Damasio, A.R., and Maurer, R.G. (1981). Gait disturbances in patients with autistic behavior: a preliminary study. *Arch. Neurol.* 38: 646–649.

Vogt, C., and Vogt, O. (1919). Ergebnisse unserer Hirnforschung [Results of our brain research]. *Jounrnal Fur Psychologie und Neurologie 25*: 277–462.

Vogt, C., and Vogt, O. (1926). Die vergleichend-architektonische und die vergleichend-reizphysiologische Felderung der Grosshirnrinde unter besonderer Berucksichtigung der menschlichen [The comparative architectonic and physiologic divisions of the cerebral cortex with particular emphasis on the human]. *Naturwissenchaften* 14: 1190–1194.

Volkmar, F., Chawarska, K., and Klin, A. (2005). Autism in infancy and early childhood. *Ann. Rev. Psychol.* 56: 315–336.

Von Hooff, J.A.R.A.M. (1962). Facial expression in higher primates. *Symp. Zool. Soc. Lond.* 8: 97–125.

Von Hooff, J.A.R.A.M. (1972). A comparative approach to the phylogeny of laughter and smiling. In: *Non verbal communication.* Hind, R.A. (Ed.). Cambridge, UK: Cambridge University Press, pp. 209–241.

Walshe, F. (1935). On the "syndrome of the premotor cortex" (Fulton) and the definition of the terms "premotor" and "motor": with consideration of Jackson's views on the cortical representation of movements. *Brain* 58: 49–80.

Weinrich, M., and Wise, S.P. (1982). The premotor cortex of the monkey. *J. Neurosci.* 2: 1329–1345.

Weinrich, M., Wise, S.P., and Mauritz, K.H. (1984). A neurophysiological study of the premotor cortex in the rhesus monkey. *Brain* 107: 385–414.

Williams, J.H., Whiten, A., Suddendorf, T., and Perrett, D. (2001). I. Imitation, mirror neurons and autism. *Neurosci. Biobehav. Rev.* 25: 287–295.

Wise, S.P., Boussaoud, D., Johnson, P.B., and Caminiti, R. (1997). Premotor and parietal cortex: corticocortical connectivity and combinatorial computations. *Ann. Rev. Neurosci. 20*: 25–42.

Wise, S.P., Weinrich, M., and Mauritz, K.H. (1983). Motor aspects of cue-related neuronal activity in premotor cortex of the rhesus monkey. *Brain Res. 260*: 301–305.

Wolpert, D.M., Ghahramani, Z., and Jordan, M.I. (1995). An internal model for sensorimotor integration. *Science 269*: 1880–1882.

Woolsey, C.N., Settlage, P.H., Meyer, D.R., Sencer, W., Hamuy, T.P., and Travis, A.M. (1952). Pattern of localization in precentral and "supplementary" motor areas and their relation to the concept of a premotor area. In: *Association for Research in Nervous and Mental Disease*, Vol. 30. New York: Raven Press, pp. 238–264.

Yeomans, J.S., Li, L., Scott, B.W., and Frankland, P.W. (2002). Tactile, acoustic and vestibular systems sum to elicit the startle reflex. *Neurosci. Biobehav. Rev. 26*: 1–11.

Zhang, T., Heuer, H.W., and Britten, K.H. (2004). Parietal area VIP neuronal responses to heading stimuli are encoded in head-centered coordinates. *Neuron 42*: 993–1001.

Index

action categories, 154–55. *See also under* stimulation-evoked movements
  ethologically relevant, 154, 160
action modes (of monkeys), 140–43
  switching among different, 141
actions, defensive. *See* defensive behavior
action zones, 5, 51
active spread (neuronal signals), 95
Asanuma, H., 40–41, 91
autism (and motor control), 192–94
  cellular abnormalities in, 194–95
  extreme male brain in, 196
  high *vs.* low level in motor control, 197
  mirror neurons and, 61, 195
  motor deficits in, 192–94
  motor hypothesis of, 194, 195
autism syndrome, 192

Beevor, C., 22–24
Betz cells, 27–28
"biblical cells," 118
bicuculline, 117
bilateral coordination, 69
bimodal neurons, 56–57
body schema, 175–76
Broca, Pierre-Paul, 14–15
Broca's area, 16, 61
Bucy, P. C., 29, 68

Campbell, A. W., 26–28, 52
cerebral cortex topography, 151
  why it is organized according to proximity, 151
Cheney, P. D., 41, 42, 73, 92
chimpanzee brain, 24, 25
cingulate motor areas, 12
cingulate sulcus, 163
climbing, properties of, 69
climbing postures, 120
"come here" gesture, 190
convulsions, 14, 16, 17
Cooke, Dylan, 139

cortical motor system in monkey, divisions of, 9
  proposed hierarchy of, 51–52, 65
cortical-spinal-muscle system (CSM model), 167, 169, 175.
  See also feedback; muscle synergies
  effects of cortical stimulation and, 177–80
  goals of building, 167
  simplified schematic of, 177–78
cortico-muscle connectivity modulated by proprioceptive feedback, 110, 111
crying, defensive reactions and, 187–88
cytochrome oxidase, 54–55

defensive behavior, 146–48
defensive movements, 59, 66
defensive reactions, 66
  components, 182–83
  divergence into social behaviors, 190
  social displays and, 181–90
  types of, 182
defensive zone, 160–61
dimensionality reduction, 152, 155–56.
  See also self-organizing map model of motor cortex
direction tuning, 77–80. *See also* tuning
  local and global, 127–30
direct spread (neuronal signals), 95
dog brain, 17–20
Donoghue, J. P., 45
dorsal premotor area (PMD), 9, 11–12, 62–63
Dum, R. P., 64

eight-dimensional (8-D) posture space, 130, 133–34
electrical stimulation. *See* stimulation
epilepsy, 14, 16, 17, 31
Ethier, C., 82
Evarts, E. V., 72
evoked movements. *See* stimulation-evoked movements

evolution, 68, 164
    of crying, 187
    Darwin's theory of, 18, 24
    of defensive reactions into social displays, 181–82, 184–87, 189, 190
    of grips, 143
    levels of, 18, 24
    of the smile, 184
    stages in animal, 17–18
    of tickle-evoked laughter, 185–87
evolutionary tree of defensive reactions and social offshoots, 190
exploratory gaze. *See* gaze, exploratory
extrapyramidal tract, 31
eye movement, 161–62
eye movement studies, 87–89

feedback, movement control without, 176–77
feedback control, λ model and, 169–71
feedback remapping, 10, 72–75, 168. *See also* muscle synergies
feed-forward control and muscle synergies, 171–73, 177, 179, 180
Ferrier, D., 20–22
Fetz, E. E., 41, 92
flight zone, 188
Foerster, O., 31–32, 34
Fritsch, Frau, 18–21, 85
frontal eye field (FEF), 3, 9, 12, 21, 53–54, 88–89, 161
Fulton, J., 30–31, 35

Galvani, L., 85
gaze, exploratory, 141–42
gaze control and autism, 195–96
Georgopoulos, A. P., 77, 78, 81
gestures, social
    common arm movements and, 190–92
gesturing during speech, 191–92
"goodbye" gesture, 190
gorillas, 24
Gould, H. J., III, 44–45
grasp actions, 59–60, 145, 189. *See also* reaching-to-grasp
grimace, 183
grip, 143–45
Gross, Charlie, 57

hand
    smooth speed profile of, 103, 105–6
    some stimulation sites compensate for weight on, 108–9

hand areas, multiple, 55–56
hand in lower space, 119
hand location, 154–56
hand movements. *See also* stimulation-evoked movements
    progression of spatial locations to which they are directed, 6, 7
hand muscles. *See* somatotopic overlap
hand positions arranged across cortical surface, 103, 104
hand positions in spontaneous behavior, 148–50
hand-to-mouth interaction, 80–81, 141, 142
hand-to-mouth movements, 113–15
hand-to-mouth zone, 67, 68, 144, 145
Hediger, H., 146, 188
"hello" gesture, 190
hierarchy, 37
Hines, M., 30
Hitzig, E., 18–22, 85
homonculus, 8, 32–34
Horsley, V., 22–24
hypothalamic stimulation, 86–87

imaging of human motor cortex, 53–54
indirect spread (neuronal signals), 95
intermediate precentral cortex, 26–27
interneurons
    propriospinal, 171, 172, 178
    spinal, 172–73
interpretive cortex, 86

Jackson, John Houghlings, 15–18

Kohenen method, 155–56
Kwan, H. C., 43–44

Largus, Scriobonius, 85
lateral intraparietal area (LIP), 21
laughter, defensive reactions and, 184–88
leaping postures, 120
Lemon, R. N., 74
limb, initial position of
    and evoked muscle activity, 110, 112
local smoothness, optimization of, 152, 164
locomotion, 65–69, 120–21, 141–42, 145–46

manipulation movements, 118–19, 141, 142
manipulation zone, 67, 68, 144–45, 163

Index

map model. *See* self-organizing map model of motor cortex
Matelli, M., 54–55
medial anterior cortex, 65
microstimulation, intracortical, 39–41, 43. *See also* stimulation
midbrain, 14, 37
mirror neuron network, 60
mirror neurons
   autism and, 61, 195
   perception of actions of others and, 60–61
model motor cortex. *See* self-organizing map model of motor cortex
monkey brain. *See also specific topics*
   integration of vision, touch, and movement in, 3–7
Moore, Tirin, 3–4, 93
motor control. *See* movement control
motor cortex
   definitions and meanings, 12, 26
   discovery, 18–22
   division
      into posterior and anterior regions, 26–28, 34–36
      into primary and premotor cortex, 33–34, 62
      into three fields, 28–29
   explanations for heterogeneity in, 49
   "muscles" *vs.* "movement" view of, 23
   nested organization, 45, 46
   proposed principles to explain properties of, 6
      mechanism of movement control, 10
      topographic organization, 6–10
   purpose, 10
   as roster of separated body parts, 32–33
   theoretical framework for understanding, 10
   zones, 45
motor homunculus. *See* homunculus
motor map(s)
   elaboration of
      by Beevor and Horsley, 22–24
      by Penfield, 32
      by Sherrington, 24–26
      Vogt and Vogt and, 28–30
   interpretive nature of, 36–37
   proposal of two distinct, 26–28
motor neurons. *See* neurons
movement control, 19–20
   levels of, 17–18
   mechanism of, 10
movements. *See also specific topics*
   primary, secondary, and tertiary, 22–23
"movement" view of motor cortex, 23
muscimol, 59, 117
muscle control, λ model for, 169–70
   integrating muscle synergies with, 173–75
   limitations, 170–71
   simplified schematic of, 169, 170
muscles, pathways from cortex to, 71–75
"muscles" view of motor cortex, 23
muscle synergies, 167
   feed-forward control and, 171–73
   integrating λ model with, 173–75

neural-network models, 10
neuronal control
   of complex fragments of behavior, 77
   of movement, models of, 168–75
neuronal system, features of, 167–68
neurons. *See also specific topics*
   α motor, 169
   in motor cortex, cause useful movements, 136–37

oculomotor system, 87–89
Olds, J., 86
optimization of local smoothness, principle of, 152, 164
orangutans, 24

parietal areas, 51
Park, M. C., 45
passive spread (neuronal signals), 95
Penfield, Wilder, 32, 35–37, 52, 63, 64, 85–86
   contributions to motor cortex physiology, 32–35
peripersonal space, 147–48
personal space and defensive reactions, 188
PMD. *See* dorsal premotor area
PMV. *See* ventral premotor area
polysensory zone (PZ), 12, 57, 58, 116–18, 123
population code, 77
   based on fragments of behavior, 76–77
   generalized, 81–83
population (coding) hypothesis, 81–82

posterior parietal cortex, 122
posterior strip, 27
postural controllers, 175
  beyond, 175–76
postural control mechanism, 137
precentral cortex, 26–27
precentral gyrus, 15, 20, 22, 25, 28, 34, 115
  polysensory zone in, 58
premotor areas. *See also* dorsal premotor area; ventral premotor area
  interpreting the zoo of, 65–67
    emergent hierarchies, 67–68
    more than one function for each cortical zone, 68–69
  medial, 63–65
premotor cortex, 48, 53. *See also* ventral premotor cortex; *specific topics*
  defining, 3, 30–31
  denied by Penfield and Woolsey, 52–53
  dorsal, 62–63
  doubting, 32–35
  fractionation into separate fields/zones, 54–55. *See also under* ventral premotor cortex
  lateral, 11
premotor syndrome, 30, 32
preSMA, 12
primary motor cortex, 11, 27, 32, 48, 49. *See also under* motor cortex, division; *specific topics*
proprioceptive feedback, cortico-muscle connectivity modulated by, 110, 111
propriospinal interneurons, 171, 172, 178
propriospinal neurons, 174
"purposive" movements, 22
pyramidal tract, 31

Rathelot, J. A., 42
reach and inward scoop, 145
reaching-to-grasp, 69, 141, 142, 145.
  See also grasp actions
Richter, C., 30
Rizzolatti, G., 54–57, 59
Robinson, D. A., 87–88
Roland, P. E., 53
rostral premotor areas, 51

sado-masochism, 189
Sanes, J. N., 74
Seidemann, E., 88–89
seizures, 14, 16, 17
self-organizing map model of motor cortex
  final state of, 156–58
  hand locations associated with categories of movement in, 155–56
  initial state of, 153, 154
  limitations of, 163–65
  monkey motor cortex compared with, 156–63
  qualitative description of, 153
    ethologically relevant action category, 154, 160
    hand location, 154–55
    optimization, 155–56
    somatotopy, 153–54
  results, 156–63
sensory perception, electrical stimulation and, 89–91
sex and suppression of defensive reactions, 188–89
Sherrington, C. S., 24–26
signal spread, 95
simculus, 35–36
single neurons (in motor cortex)
  properties of, 37
    reasons for mismatch between stimulation effects and, 125–26
  and their contributions to movement, 72, 75–76. *See also* direction tuning
    extrinsic *vs.* intrinsic coordinates, 79
    generalized population code, 81–83
    normal behavior requires control of combinations of variables, 80–81
    population code based on fragments of behavior, 76–77
    tuning to many movement variables, 79–80
single neuron specificity, 77
smile, defensive reactions and, 183–84
social displays. *See* defensive reactions; gestures
somatotopic map of brain, 31–32
somatotopic overlap
  beyond the fingers, 43
    core-surround organization, 43–44
    lack of overlap among body segments in primary motor cortex, 46–47
    maps of hand and arm representations, 44–46
  hand muscles as overlapped in primary motor cortex, 39
    motor cortex neurons correlated with many muscles, 41–42

overlapping representations of fingers, 42–43
proposal of cortical columns for individual muscles, 39–41
increase in overlap with experience, 47
as reflecting hierarchy *vs.* behavioral repertoire, 48–49
somatotopy, 37, 153–54
space, personal
defensive reactions and, 188
speech perception, motor hypothesis of, 60–61
spike-triggered averaging, 73–74
spinal cord. *See also* cortical-spinal-muscle system (CSM model)
electrical stimulation of, 121
projection systems from cortex to, 31, 48
spinal interneurons. *See also* propriospinal interneurons
how movement is controlled by, 172–73
spontaneous behavior, repertoire of postures in, 148–50
stimulation, electrical, 85, 92, 123. *See also* microstimulation; stimulation-evoked movements
artificiality of long, 93–94
of cortex, 19, 121–23
damage caused by, 94
direct *vs.* alternating current, 21
eye movement studies, 87–89
hypothalamic, 86–87
of midbrain, 120–21
postures evoked by
match between postures preferred by neurons and, 134–36
vs. postures in spontaneous behavior, 148–50
pulse amplitude, 100–101
pulse frequency, 100
pulse train duration, 101
pulse width, 99–100
shorter *vs.* longer periods of, 21
vs. single-neuron recording, 95
of spinal cord, 121
to study perception, 89–91
surface, 85–86
traditional studies in motor cortex, 91–92
stimulation controversy, 93–95
activating widespread networks, 95
biphasic *vs.* negative pulses, 93
connections *vs.* function, 94–95

stimulation-evoked movements, 21–25, 97–99, 123. *See also* monkey brain
common categories of, 113–20
in dog brain, 17–20
maps, 19, 20, 23, 25
match between natural neuronal properties and, 125–27, 136–37
optimizing for evoked behavior, 101
pathways through which they are evoked, 68
qualitative description of action categories, 113. *See also* manipulation movements
climbing/leaping, 120
defensive movements, 115–18
hand in lower space, 119
hand-to-mouth movements, 113–15
mouth movements, 119–20
reach-to-grasp movements, 119
quantitative description of convergence, 101–3
hand positions arranged across cortical surface, 103, 104
joints interaction stabilizes hand for some stimulation sites, 106–8
patterns of muscle activity, 109–13
smooth speed profile of hand, 103, 105–6
some stimulation sites compensate for weight on hand, 108–9
stimulus-triggered averaging, 42, 73–74
Strick, P. L., 56, 64
supplementary eye field (SEF), 3, 9, 64, 161, 162
supplementary motor area (SMA), 12, 34, 48, 53
hypotheses for the function of, 63–65
illustrations of, 9, 35
stimulation of, 64–67, 69, 120, 146
surface mapping, beyond, 37
Swedenborg, Emanuel, 13–14
sympathetic defensive reactions, 189

Tan, case of, 14–15
Teitelbuam, O., 193
threshold current, 100
defined, 100
threshold stimulation, 40–41, 44, 45
tickling and tickle-evoked laughter, 185–88
"tip-of-the-iceberg" hypothesis, 42, 44

Todd, Robert Bentley, 14
tuning. *See also* direction tuning
　in 8-D posture space, 130–33
　in posture subspaces, 133–34

ventral intraparietal area (VIP), 122–23
ventral premotor area (PMV), 9, 11, 12
ventral premotor cortex
　mirror neurons in F5, 60–62
　multiple hand areas, 55–56
　specialized properties of F4, 56–59
　specialized properties of F5, 59–60
Vogt, Cecile, 28, 29, 34, 68
Vogt, Oskar, 28, 29, 34, 68

Walshe, F., 32
Welch, K., 34, 63, 64
Wise, S. P., 62
Woolsey, C. N., 35–37, 52, 153, 154
wrist, muscle force at, 72–73